The Big Full Circle
The Moon's Changed And Lasting Light

Don Beeton

Copyright © 2017 Donald G. Beeton

All rights reserved.

ISBN: 1544793111
ISBN-13: 978-1544793115

DEDICATION

Very special thanks to my Family, my wonderful Sisters
and especially my very special Mom and Dad for all their love.
I want to thank my two children for all of their help and encouragement.
Thanks for everything! Especially all your love!!
Love Dad

CONTENTS

	Acknowledgments	i
1	Denying The Illusion	Pg#1
2	Many Early Hominid Types It Only Makes Sense	Pg #7
3	The Shadow	Pg #17
4	Naming Rules What Naming Rules	Pg #40
5	Countenance," It's Shocking But This Is What Happens	Pg #48
6	The Big Full Circle	Pg #58
7	The N.L.T.	Pg #65
8	Nobody Lives There	Pg #85
9	The Shape Effects	Pg #100
10	The Return Viewer's Guide	Pg #114

ACKNOWLEDGMENTS

Back Cover Art: *The Crossing Down*
Copyright © 2016 Sandra Beeton
All rights reserved.

My daughter surprised me with this wonderful gift. As far as I know this painting is the only painting in existence at this point in time ancient or modern that clearly shows the main elements involved in a Return. Light from outer space, the Earth, the Moon and the moon coloured Ancient Object of the Crossing Down itself complete with the Mound, the Shadow and the Standing Stone Craters.

Front Cover Images
Nasa Images: *I would like to thank both Buzz Aldrin and Neil Armstrong.*
Buzz Aldrin photographed by Neil Armstrong, Apollo 11
Brooklyn Museum Archives:*Lantern Slide Collection, Great Sphinx, Gizeh*

I'm not a writer, that explains a lot. I have published without proper professional book editing. I need an editor. I continue to work finding and fixing my many errors overwriting online files. I have to try to write about things that should be impossible in a way that is clear so that people can start to understand the important things that I saw. My lack of writing experience makes this task difficult for me and difficult for the reader. Patience is required. I'm hoping that people will remember that the bottom line for me is to publish my information to try to do the right thing for my children, my family and for everyone else and for myself as well. I know that what I saw is very profoundly important. I want to thank everyone who reads my descriptions and viewing advice. I hope it turns out that you get to see the Object and you get to see in the ancient way from within the Changed and Lasting Light from the Moon. I express my opinions and guesses about various topics. I'm certainly not always right about things and sometimes my opinion on various questions changes as I discover and learn more. I believe it's important for you to see the Man of Light because this is your own personal human heritage. Everyone has equal responsibility to help another person to see this especially so for parents who all have added responsibility to try to ensure that their children know what happens and that they are given the best possible chance to see and know the face of the Man of Light.
My content and more is available free for everyone to read and know
returnviewersguide.ca

1 DENYING THE ILLUSION

That's what I did.

I tried to down play the Lasting Light Event as a way to be able to report the fact that I had seen the Ancient Object, this huge place that was just simply suddenly there coming down and rolling by.

Just to start with the sight of this massive crater covered forward rolling celestial object seemingly basically headed in your direction is easily a completely heart stopping and heart accelerating sight to see all by itself before the Lasting Light event even starts! A very scary big and very important thing to know about! This massive place does roll down past us every forty plus years or so and I know that it is as simple as that if I'm right about it's 46 year and one day long orbital time period..

I saw the brilliant hard rolling ground of the surface of the Ancient Object. As a kid I thought that except for the way it's craters all looked with the standing crater rim pieces that circle every crater the Object looked like the Moon all cleaned up swept and vacuumed and even polished. Certain things are simple for me. I know that the Object I saw is real and it is headed back basically right now. This is simple but actually very difficult to know. Then there is the sight of the Moon's changed and Lasting Light, what I used to always call "the illusion," a very very difficult thing to know. Not only that but when it comes to talking and writing about what I saw especially what happens within the Moon's changed light the difficulties I know and feel increase.

How do I report the orbiting celestial Object I saw and what I perceive to be a potential dangerous situation caused by the passing of the Object while at the same time there is this fantastic bizarre and wonderful thing that it does the Changed and Lasting Light event from the Moon? The Awesome Good the sight of the ancient muscular Man of Light really does happen there and also much much more! The whole Ancient Object and Changed Light business and everything that happens to the viewer has never been easy for me to think about. For most of my life it was easier to think about the Object itself as opposed to the Changed and Lasting Light from the Moon.

Who do I report what I saw too?

In an early effort to report my sighting of the crossing Ancient Object I figured maybe I could basically stick to the Object itself and it's flight path and what the craters looked like etc, at least as my focus. I figured that if the Object I saw and knew was known in some circles descriptions of the Ancient Object's surface itself would be enough and the fact that I had seen the Object would be obvious.

I eventually started to really think that I could possibly successfully report the Object if I focused on the Object instead of what the Object actually does when it effects the sight of the Moon's light when it changes the Moon's light and the light from the background of space.

This lead to a completely misguided and unsuccessful effort.

The experience of this failure caused me to except the sights that I saw much more easily. I started to learn to except the things I saw and the things that I know. Also it's been a long process for me to finally realize that the Object I saw really is unknown. Even a person who I thought would know if anybody actually knew clearly did not seem to actually know about the incredible ancient astronomical object that I saw.

I reported my sighting of the Object to a person that I have a great deal of respect for and I still do have great respect for this person. In many ways I was starting out at that point and I was not prepared to really face the sight of the Awesome Good once again. I had good intentions but I was not used to describing my view and I was not really willing to just simply say the things I knew needed to be said. Even though I knew a Return is about the Ancient Object and the changed light from the Moon I was mainly focused on the hard ground of the Object itself and at that time I would have preferred to keep things that way. Once I did mention the Changed Light I immediately regretted mentioning the Changed Light and I back tracked trying to suggest that the light that came from and then began to rise from over the Object's upper right horizon was basically just to bizarre and that I had been badly fooled.

Looking back I am actually happy with the way everything turned out. If that credible person had shown any interest I don't think I would have tried as hard. Definitely I know I would not have been were I am now.
I would like to thank this person for showing no interest. Finally I got to work.

It was a shock for me after I had spent even years deciding I had to do some thing and then months bracing to finally click send in order to begin to report my view. Then a huge shock for me! No interest! I did get a response from this person but there was no interest That was a major blow for me, one that I now know I needed.

What should I do now?
In my mind I had just gone to the top and it had got me no where! Real shock and then gradually more and more a feeling of real desperation started to grow. There are some private and personal aspects to this as well that I will not be discussing. To try to quickly sum this up and get past this point I learned to focus on this and I learned how to focus on the very difficult even intimidating and very unnaturally fresh memories of the Changed Light. Out of my desperation and eventually very directly out of my determination suddenly a few weeks later or some time later the most incredible thing just simply even magically happened!

Suddenly as plain as day The Large Shape Effect was just simply right there for me to see! All of a sudden I realized that I had been seeing it in a peripheral sort of way on and off for a number of days but now suddenly there it was, I was seeing it at the same time I was thinking about it and this caused me to suddenly realize The Large Shape Effect!
I already knew this after glow shape but hadn't really seen it this way for many years. Fantastic! The actual greatest thing I know!

And there falling away from the Valley was the actual rolling ground of the Ancient Object itself down low actually rolling and turning at the bottom of The Large Shape Effect. Along with the whole Large Shape Effect itself, I had suddenly rediscovered the hard rolling ground of the Object itself!

I suddenly was actually able to see the horizon! The place were the Object ends and the Changed Light from the Moon starts. The actual shape of things!

It turns out that this is huge! This is a very important basic thing to know!I am very convinced that in the long run it will be the idea and the mystery and the awareness of the existence of The Shape Effects and the imprinting effect that the Changed Light has on the viewer that will be the most important thing or contribution that I will have made.

I had it all backwards before when I was trying to approach a single individual with my descriptions of the Ancient Object I saw while trying to avoid the Changed Light and even backtracking and wishing I hadn't mentioned the Changed Light after I did mention the Changed Light. Now I understand the point is the Object yes that's true but most importantly the fantastic thing is what the Object does and not the Object itself.

The point is to forget about one guy even if he is at the top and instead tell everyone about the Changed Light and what it looks like and what it does and then try to describe the fantastic part about The Shape Effects, the actual after effects that the Changed Light leaves you with!

People need to know about this! I can't concern myself with single individuals even if they are at the top. Eventually the people at the top are going to catch on and they will in the mean time the point is for people to have heard the idea of the Ancient Object and the Changed Light event that it causes. The Awesome Good, the sight of the muscular Man of Light that does happen there and ultimately the imprinting after effects of seeing and feeling the intense pressure of the Changed Light, The Shape Effects, that's the point!

In the end after all the tough questions it's the incredible Shape Effects that are at the very core of the big point of it all! We are exact copies of what the very heart of the Changed Light looks like. The imprinting effect that the Changed Light causes to happen to the life that sees it somehow causes some of that life to become the same shape and form. This is at the heart of what I understand to be the actual ancient mystery itself. Our ancestors knew this and realized the importance of The Shape Effects.

I've gone from trying to report the Object at the expense of the Changed

Light to focusing on the Changed Light and describing what it looks like as much as I talk about and describe the Object itself. The Object is beyond awesome and incredible and impressive and profoundly important but the Moon's Changed and Lasting Light really is the big ancient important thing that somehow caused us to be here! I feel much better now that I finally understand some of the big main point of it all.

Being able to later see the Large Shape Effect and The Small Shape Effect and the reversible movie clips and the very unnaturally clear memories and everything that has to do with the qualities of The Shape Effects are at the very heart of the ultimate ancient mystery!

As I try to describe this is going to the most basic level and is at the very heart of the Awesome Good and the reality of human life in it's present form! This truly is a key to understanding.

This is something that people in the future will be intensely studying and deeply involved in at a far deeper level than I am able to go. If I can help people to realize The Shape Effects are going to automatically happen to them when they see the Changed Light then that will actually be a very real contribution.

My Return Viewer's Guide is obviously important and at the end of the Black Road is the big ancient sight were we come from but beyond is the effects and the influences of The Shape Effects. The imprinting effects of the Shape Effects are actually somehow at the very heart of us all!

It's the power and the pressure and the screaming intensity and the feeling of the sight of the Changed Light focused on you and it's after glow effects the way the Changed Light sights imprint themselves into you that is a big part of the Awesome Good itself! I am very lucky to actually know this!

If the individual who I approached had been interested I may not have felt the desperation and I may never have seen The Shape Effects again. I am very fortunate that this individual was not interested.

Dear sir,

If you ever find yourself reading here on this page I really have to thank you for your complete and total lack of interest and your "lack of expertise in this area," as you put it.

Thanks, you helped me forward. I don't really know what I expected from you. E-mailing you was like making a commitment to do something about this finally. Thanks for being polite. Also once again thanks for doing the important work that you do.

I hope that you will also have a chance to see the Object of the Crossing Down and especially the incomparable special Changed and Lasting Light from the Moon.

all the best.

Very sincerely
Don Beeton

2 MANY EARLY HOMINID TYPES
IT ONLY MAKES SENSE

Something very massively profound caused Homo Sapiens to exist. Obviously everyone knows this already. Surely like so many other things we will never know the entire story of what exactly caused humans to exist. We as humans have many ideas and theories that try to answer this difficult question. Obviously everyone knows this as well.

Certainly as a young boy I had no idea what the sight of the Man of Light was all about and I still don't know anywhere near the whole story and I don't think anyone ever will. I have had many thoughts on this subject ever since I saw the big main black shadow strike and draw the squiggly line that leads to those seconds where you find yourself looking down at the muscular back and the right side of the Man of Light from above and behind. It's been a long road trying to begin to understand.

There are many places on my website where I talk about some of the difficulties and problems I had from those early days going forward firstly trying to understand what had happened to me when I saw the Lasting Light and what was the meaning of what I saw and what is the Changed Light or rather how does the Changed Light actually happen?

There is a Big Full Circle that at the very least the idea of which is a piece of the puzzle. Obviously this reflects my personal opinion and I don't expect people to automatically agree with me. I does seem obvious to me that there is a basic question that a person is confronted with if they are fortunate enough to see the sight of the Man of Light.

Does the sight of the Man of Light up there have any connect with us down here? This is a fundamental question that will simply present itself and every person who sees will know first hand this question is unignorable and has to be asked. All areas of study and human understanding have to be included in the question and in trying to answer this question. To accidentally or purposely leave any area of science or religion out of the question is to miss or ignore the basic question that the existence of the sight of the Man of Light forces us to ask. Does the sight of the Man of Light have any connect with at all with us? There will be many answers given and reasons given for those answers. And also many opinions regarding what it all means will be expressed and of course these opinions will be all equally valid and correct.

Answer the question: Does the sight of the Man of Light have any connect with at all with us Homo Sapiens? If you think it does many people will agree with you. Answer the question no and you will also have many people who agree with you. Both sides of the question will give their reasons. At this point hopefully people will figure out that it does not matter if a person sees this question differently than they do. It actually makes no difference what the difference of opinion is in the end and it comes back to automatically respecting that other person that you either know or don't know. We are all humans and we all deserve automatic respect from each other. In my opinion this basic human value and right and it is part of what the sight of the Man of Light means to the discussion and to the human condition.

We are life forms that are exact copies of what the main most important central element of the first part of the Lasting Light event looks like! We are self aware and we are not just capable of loving each other we are capable of loving other life forms as well. This is a fantastic fact! In my opinion this points straight towards the heart of the ultimate ancient human mystery.

To more fully point towards the ancient mystery I think that clearly the mystery of the Changed Light event that the ancient crossing Object causes to occur is the first consideration at the root of it all. Life somehow started on the Earth and eventually some of that life looked up and started to see the Changed and Lasting Light including the human form. This is directly related to the imprinting effects that the Shape Effects causes to happen to the person or life form that happens to be

MANY EARLY HOMINID TYPES IT ONLY MAKES SENSE

seeing the Changed Light shapes and sights. Somehow we are exact copies of the heart of the Changed Light. My guess is that just by looking at the muscular human form of the Man of Light our earliest ancestors were caused to grow and evolve into that shape. Not only were we caused to come into existence but it appears as though many hominid types were also caused to rise and come into existence.

All the special sights that are seen other than the crossing ancient Object itself are sights that are in fact light from the Moon and the background of space that has been effected by the Object's forces. As if a statue down in front of you the first sight you see after the big main shadow strikes is the crystal clear sight of the very muscular back arms and shoulders and right side of a man sitting or crouching in the thinker position looking away form you directly towards the distance of the Valley. It may be possible that this figure is not sitting in the classic thinker position even though I describe him that way. This is were and how it all starts as far as the first sight of the human form is concerned.

I wrote this on my website home page:
"The crossing forward rolling speeding Ancient Object is beyond stunning and spectacular but somehow it turns out that it's all about the point of a shadow and the fantastically changed incredible spectacular light from the Moon."

Incredibly there you are watching an awesome celestial object rolling at you from in front of the Moon and then suddenly after watching and following the point of the big main shadow over the valley floor layer of flat white clouds into the extreme depth of the valley to the upper right you are suddenly looking at the human form in extreme muscular detail!!! Fantastic and mind bendingly bizarre but incredibly that's what it is and that is what happens!

Suddenly you are also noticing that way the Changed Light and Lasting Light causes you to see is the the same as if out of a remote camera up there. It's all about the Casting Light and what it looks like and how you see from within it and from whatever place you look at. Look at a place in the Changed light and then you will be suddenly looking at that place from up close. Then at this point you can look in a different direction and see in that direction from that place. This effect first starts when the Object's surface suddenly looks flat and you realize you are seeing the surface from a very close distance above it. At this point look to the right and see the suddenly formed ancient Valley.

It's the Moon's suddenly visible light and somehow it has become drastically changed in appearance and behavior by the Object's powerful forces. From this early point onwards from the outer edges of the Lasting Light inwards the light is imprinting itself into the viewer. I call it the Shape Effects. Part of this effect is the intense afterglow that stays with the viewer after seeing the Changed Light. Every future viewer will automatically have this after effect happen to them just by looking at the Changed Light

Whatever effect that causes this imprinting into the viewer as I describe is at the heart of what I think of as the ancient mystery and the process that caused some of the life that saw the Man of Light to be caused to grow into and eventually become an exact copy of the Man of Light. The overall shape of the Changed Light and the shapes seen within and the special qualities of the Changed and Lasting Light is at the heart of how the process worked and works causing our existence along with many other hominid types throughout the past. I can speculate and wonder and ask questions and I have opinions. This doesn't mean I'm right about things. It's up to everyone else to also ask their own questions for themselves. It will be up to you to wonder about this as well. Check to Return Viewer Guides and prepare to see so that you have the best chance possible to see the Lasting Light sights that are seen. Either the sight of the muscular Man of Light is connected with and responsible for humans or there is no connection at all and the Man of Light and us humans on Earth exist separate and apart from the other with no connection whatsoever, period. It's one situation or the other and beyond that I can't see a third answer or possible third option pick from.

The big questions will still be there. It will matter but it also won't matter to people what the process or the natural mechanism was that created us humans. Many people will believe there's a god behind all of this and they will have faith. No one will be able to disprove them in their belief. The sight of the Lasting Light will be their proof. Others will see this as an act of nature and we resulted without a god type being involved. They will consider the sight of the Lasting Light to be their proof. No one will be able to disprove this interpretation either. In the future as in our time it seems to me that certain questions will simply remain unanswerable.

From Wikipedia the genus of hominids:
"Current evidence indicates that there were as many as 12 species of early hominids between 6 and 1.5 million years ago, but they did not all live at the same time.

MANY EARLY HOMINID TYPES IT ONLY MAKES SENSE

The following species are the most widely accepted ones:
1. Australopithecus anamensis
2. Australopithecus afarensis
3. Australopithecus africanus
4. Paranthropus aethiopicus (or Australopithecus aethiopicus)
5. Paranthropus boisei (or Australopithecus boisei)
6. Paranthropus robustus (or Australopithecus robustus)" wiki

Not only were there many types of early hominids but the great apes and all primates are clearly also a central part of the mystery. The basic musculature of all the apes as well as humans is the same with minor anatomical differences. When you look down from above and behind and you see the muscular back of the Man of Light you see the human form very clearly and a muscular back and shoulders and right side look they way they do and that's exactly what this incredible sight looks like.

"The Hominidae, whose members are known as great apes or hominids. The current, 21st-century meaning of "hominid" refers to all the great apes including humans.

A hominid is a member of the family Hominidae, the great apes: orangutangs, gorillas, chimpanzees, and humans.

A human is a member of the genus Homo, of which Homo sapiens is the only extant species, and within that Homo sapiens sapiens is the only surviving subspecies." en.wikipedia.org/wiki/Hominidae

Elongated Skulls Copied Human Looking Shapes Seen In The Moon's Changed Light

The compelling elongated skull evidence that has been uncovered and brought to light has caused researchers to ask a startling but logical and obvious question:

Do the elongated skulls in particular the elongated skulls from the Paracas Peru region represent a genetically different line of hominids compared to homo sapiens?

When I first heard about the stunning evidence and facts that clearly indicate that the answer to this question might be yes I knew that the researchers looking into this question might be correct in their conclusions. I realize that it might be possible that because of the way that homo sapiens are an exact copy of one of the shapes that is seen in the Moon's Changed Light event that the Crossing Object and it's shadows causes in and with the light from the Moon there may be the exact same situation in place in regards to human looking shapes with elongated skulls created by shadows within the Changed Light.

Did a natural evolutionary line of hominids arise that featured elongated skulls the same way we homo sapiens arose and evolved from looking at a Changed and Lasting Light human form that has a skull like our skull?

It appears that there are many examples of elongated human skulls from many places around the world and I have seen pictures of some examples over the years. Elongated heads are shown in ancient depictions. Apparently the practice of cranial deformation was developed and widely used around the world to cause a child's skull to become permanently elongated as that person matured. Until now I had always thought that this practice accounted for all of the examples of elongated skulls as it never occurred to me think or wonder otherwise.

I always thought that people practiced skull deformation as a way for humans to copy certain human forms seen within the Moon's Changed and Lasting Light. Beyond that point I didn't wonder if there was anything else or other factors to consider that were part of the elongated heads story.

Neither of the two types of human forms that I saw suddenly created had elongated skulls. Certainly the muscular Man of Light at the heart of the Changed Light Valley and the rising light mountain does not have an elongated shaped skull. I did not directly closely see all of the Easter Island type of human forms that suddenly appear when the Changed Light rises.

The Easter Island type of individual located to the Man of Light's right is the individual that I saw directly from close up the way the Changed Light allows you to see and this human looking shape does not have an elongated skull as far as I know. Does the round brimmed hat he wears conceal an elongated skull that will be seen later? Maybe I can only guess about things I didn't see that either do or don't happen later in the Moon's Changed Light. The scene in the Changed Light progresses and evolves and for example the Man of Light transforms and changes and the Easter Island type of shape my possibly transform in appearance at some point as well. It may be possible that some of the Easter Island type of individuals have elongated skulls however I did not see this effect. The Easter Island style of human forms that rise together at the same time the suddenly transformed looking Man of Light rises all have the same general look or group appearance as they rise in their double curving arc arrangement flanking on both sides and around in a

MANY EARLY HOMINID TYPES IT ONLY MAKES SENSE

semi circle behind the Man of Light oriented all facing towards the Object's upper right horizon from their location standing on the floor of the sunken room at the top of the rising Light Mountain. This is truly a most incredible thing to see.

I think that it is very probable that the human looking forms with elongated skulls were seen during the third phase or stage of the Changed Light event, the oval shaped plain. I say this because as far as I am able to determine there are three stages or main phases composing the Changed and Lasting Light event. I saw the entire Valley phase and a good portion of the rising Light Mountain phase and I did not see a human form that featured an elongated shaped skull. Perhaps the human forms with the elongated skulls appear not just within the area of the third phase's oval shaped plain but maybe they appear in association with or perhaps they are seen located within the area of the concentric rings that appear to have been seen created in the Changed Light.

The ancient people all seem to be describing this concentric ring effect and this sight was seen and copied and talked about by humans in legend and stone. Certainly the ancient people knew that they themselves were copies of the very muscular shape of the Man of Light at the heart of the Changed Light event. It may be possible that cranial deformation was practiced so that people, homo sapiens could change the shape of their children's skulls in an effort to copy a type of human form that they saw within the Changed Light that they obviously held in very high regard. Also it may be possible that homo sapiens practiced cranial deformation as a way to copy, imitate and emulate a different hominoid species that lived and walked among them in the world at that time.

It occurred to me to ask a question; We are an exact copy of a human form that is seen within the Moon's Changed Light. It appears as if humans and all types of hominids looked up and saw the human forms within the Changed Light and were caused to grow into and copy the human looking shape that they were looking at just by looking at it. When I saw and went face to face with the widely spaced eyes of the Man of Light I really noticed that he has a bony looking face. I have always felt that the imprinting effect that the Changed Light causes an effect I call the Shape Effects is closely tied together with why his bony face looked so natural in appearance to me. I believe that beginning to understand the Changed Light's imprinting effect that just simply happens to the viewer simply by looking begins to scratch the surface of the actual ancient process that caused us to exist in our present form. Did a natural evolutionary line of hominids arise that featured elongated skulls the same way we evolved from looking at a Changed and Lasting Light human form that has a skull like our skull?

Obviously this is a question, that people can't consider today however to me this seems like a completely obvious question to ask. Humans coexisted with Neanderthals and perhaps other hominids as well. I really do not think the full extent of the human story is known today so who knows what actually really happened in our past. To me it seems completely possible and even somewhat probable that there could have been a type of genetically separate hominid that featured an elongated skull. The next question is are the hominids with the elongated skulls if they existed responsible for the very ancient megalithic constructions that can't be explained today? Maybe this will be a question that will be answered one day.

Many researchers today believe that they see many indications of an ancient world wide disaster. I am mentioning this because the evidence for this is very compelling without question. I think they might be right in their conclusions. Perhaps the apparent cataclysm that seems to have struck the Earth and it's inhabitants around 12,000 years ago ended the hominid line that featured elongated skulls. Perhaps hominids with elongated skulls died out for a completely different reason much more recently around 2000 years ago as is suggested by researchers. Clearly this scenario is also possible. If the theorized great cataclysm of 12,000 years ago didn't happen I suspect that our world today would be a drastically different place perhaps with hominids with elongated skulls still highly positioned in society. Perhaps they might even be rulers calling the shots and running the show continuing to build fantastic megalithic constructions many of which copy exactly what they saw in the Moon's Changed Light. Specifically I believe they copied the things they saw during the third stage of the Changed Light event where everything that happened earlier from the time that the shadows started sculpting and shaping in the Changed Light was then seen again as a part of and within the oval shaped plain.

If we are lucky enough to see the oval shaped area of Changed Light emerge from the upper structure of the rising Light Mountain we could find ourselves seeing the actual sights and Lasting Light things that the ancient people were always most focused on. We'll see what happens the next time the speeding ancient Object rolls down across in front of the Moon and causes the Changed and Lasting Light event to happen again. I saw the Valley phase and the rising Light Mountain phase of the Changed Light event. I am guessing about what happens next in the Changed Light past the point that I saw up until based on the clues left by the people in our distant past. I am very certain that if the oval shaped third stage of the Changed Light event occurs during the Object's next crossing down through between the Earth and the Moon we will see many familiar looking structures and fantastic things. Look at the ancient and megalithic constructions that we are aware of today and then during the Object's crossing down in front of the

MANY EARLY HOMINID TYPES IT ONLY MAKES SENSE

Moon look and see if you see any of those shapes recreated again within the Changed and Lasting Light from the Moon.

Imagine a world where all the incredible places like Tiwanaku and Puma Punku in Bolivia were never destroyed and they still existed in original condition in our time today complete with an unbroken history and a world society of some description. A society that was focused on their true origin, the returning Object of the Crossing Down and especially on the Changed and Lasting Light event that the Object causes to happen to the light from the Moon. If they were not destroyed by a cataclysm of some sort how far would the megalithic builders have advanced in their next 12,000 years up to this point in time? If that happened we wouldn't be here to wonder who built places like Tiwanaku, Puma Punku, Gobekli Tepe and Gunung Padang and many other places instead we would have been the future that never happened. In a strange sort of way the cataclysm that destroyed the megalithic builders and perhaps also the hominids with the elongated skulls allowing our future and present time to exist is similar to the way that the extinction of the dinosaurs set the stage for the rise of mammals and ultimately all the different varieties of hominids that walked the Earth.

Life on Earth somehow arose and then some of that life looked up and saw the muscular finely detailed shape of man formed like a three dimensional statue made of solid looking unmoving very special Changed Light from the Moon. Somehow over time some of the life that looked up at this shape must have been caused to grow into the human form that was being observed in the the Changed Light. The intense imprinting effects of the Changed and Lasting Light must be central and at the heart of how this ongoing fantastic process works. This is more than just a guess that I am making it's a combination of many things. The intense pressure of the rising Changed Light mountain is certainly imprinting into you but also the imprinting effect starts from the first instant that the Moon's light seen suddenly changed. This is when the ancient Valley is first suddenly visible beside the Object to the right and beyond the Object's upper right. There's no pressure felt from the sight of the ancient Valley and also you experience no pressure from looking at the Changed Light that forms the Man of Light when he is first seen created by the big main shadow. Because the entire shape of the Valley is seen as a part of the Large Shape Effect I know that imprinting is going on or happening from the very start of the sight of the Valley even though you don't feel the Changed Light's pressure the way you are intensely subjected to a force or pressure the way that happens when you see the Changed Light mountain rising the way I describe in the Return Viewer's Guide. There is an incredible physical thing that happens to you automatically when you see the Changed Light. This effect happen to me and it will automatically happen to you when you see the ancient Object of the Crossing Down and the Changed and Lasting Light from the Moon. This process is at the heart of the actual ancient mysterious process that caused us humans to come into existence with our skulls the way we know our skulls in their present form.

How many other forms of life on Earth were caused to exist today in their present form because their shapes are seen within the Changed Light? Not just many early hominid types and possibly hominids with elongated skulls but also animals of various types might have been caused to come into existence because their ancestors saw the Changed Light and their own particular shape within it exactly the same way our ancestors saw what became their shape or our shape within the Changed and Lasting Light.

Everything always does come back to the question does the sight of the Man of Light up there have any connect with us down here? This is the fundamental question. In my opinion the answer is yes there is a connect that reaches to a deep biological level within us humans and all hominids.

It does look like a scientific fact that there were many many early hominid types in our history. After seeing the Man of Light and then through my life reading about early human history I have to say that it only makes sense to me that there were many early hominid types that may have also included hominids with elongated skulls. Fortunately the evolutionary line that led to us survived. Now it's our responsibility to future humans to make sure our evolutionary line continues to remain unbroken with a healthy planet to live on.

3 THE SHADOW

The sudden appearance of a growing very black spot on the very bright surface of the Ancient Object in the left area of the overall scene drew my attention immediately. There is a very sharp visual contrast now within the scene. Suddenly the first step in the sequence of events that leads up to the sculpting of the incredible muscular Awesome Good starts to happen.

On the left the forward rolling Object's raised mound area becomes a rounded gradually occurring overhang. Phase one of the visual spectacle of the Changed and Light from the Moon the valley has been well underway in place for a short length of time leading up to the first appearance of the black spot of the shadow.

"There came out Arukhas, solid and heavy and very black. And I saw how suitable he was. And I said to him, Come down low and become solid!" Enoch 2

The first sudden movement actually startled me. The very black spot of the shadow spreads out underneath the growing evolving overhang and is seemingly stuck up there. At first I did not recognize the shadow as being a shadow because of it's strange appearance and behavior. As it gathers in more ceiling surface area it thickens and also starts to begin to slowly spread and reach down the face of the now very steep wall like surface area that leads down to the floor of the valley. All of this occurs in the left area of the valley scene on the actual real ground of the Ancient Object's very fast moving forward rolling surface as opposed to the valley to the right. The very expansive valley is the Moon's light spread out extremely wide and deep.

The first area of the valley the wave of the shadow will travel through is the surface of the Object itself from the left to the right traveling to the upper right away from you through the craters that are evenly spread out in this area of the Object's surface towards and eventually past and over and beyond the Object's upper right horizon. All of the craters have massive towering crater rim pieces . A very high number of individual shadows are always adding to the complexity of the shadows awesome wave.

The forward rolling crossing Object eclipses the Moon and casts many individual shadows into the Moon's Changed Light. Each individual shadow has the potential to create some sort of shape or form out of Changed Light that then lasts and then would have or could have be seen later within the Moon's Changed Light.

Each individual shadow does it's own very particular sculpting and shaping and creating while building out and down with each arriving wave of Changed Light from the Moon and also Changed Light from the background of space. This goes strait towards the central reasons why phase three of the Changed Light event the oval shaped plain was said to be so fantastic. I did not see that far along into the event as my view ended. Also as I have said before it does not appear as if the Lasting Light oval shaped area of Changed Light plain part of the event occurred during the last return. Not only that but it appears as if the Changed Light that used to emerge out of the mountain of Lasting Light after being transported towards the Earth may have stopped occurring long ago back in ancient times for reasons that I can only guess at.

The sight of the valley sets the scene. The shadow's wave will ultimately move toward the valley, the Moon. The sight of the valley is blended with the Object's upper right horizon and is a stable area that does not roll with the Object. The changed but stable looking light from the Moon blends with the Object where the Object's far sloping horizon blends down as part of the valley. It all becomes one very big looking extremely expansive place.

As the Object's forward rolling motion continues the shadow finishes growing in size and thickness under the newly formed overhang and then very suddenly leaps or even pounces down to the valley floor below were it instantly starts moving across the surface of the Object from the left area of the overall scene. Instantly the shadow moves to the right at a constantly increasing angle as it starts to rapidly move across from left to right and away from you. I only finally recognized the big moving black thing as a shadow once it's awesome wave started across the Object's surface now looking the way you would expect a shadow to look if you were looking down on it from above as it crossed over land.

Somehow the shadows are the actual key or at the very least a very central part of the question of how does this happen? The shadow seems to trigger the second phase of the Changed Light event. This may or may not be true but as soon as the shadow moves off the actual surface of the Object itself into or onto the area within the valley were the flat white cloud effect is seen instantly the flat white cloud layer changes into flat smooth hard ground matching the edge of the shadow's outward movement. The shadow is already well into the great distance of the Moon's Changed Light creating new ground out of the flat white cloud effect before when shadow's central leading main shadow's point strikes. At this moment the heart of the Changed Light now sculpted and shaped with the muscular human form is suddenly gently moving on an angle upwards in your direction. This is the same place and this is at the same point in time when you are seeing from a very close position looking down from just above and behind at the back of the Man of Light.

Does the shadow actually trigger the light's movement and sudden solid appearance or was the edge of the first area of effected and changed solid looking and now Lasting Light already moving towards you over the Object and the shadow just happens to pass over this area at the exact time? The sight of the valley is the sight of a place that is just suddenly there to be seen. It is not moving towards you in any perceptible way but very suddenly it does move towards you or rather it's light is moving towards you. The shadow could be seen as and does look like some sort of trigger to the sudden way you see forms and the light moving up rising or towards you. Without any question about it the shadows do sculpt in the Moon's light. Not only that but each wave or portion of arriving one way pulsing area of light from the Moon have their own separate area or time in the area that had crater rim pieces casting shadows into it and sculpting in it.

The evolving sculpting effects are stored in the new areas of rising light that are then added into the base of the Lasting Light mountain that rises out of the former area of the sight of the valley. The light from the mountain comes from the light of the valley which is actually the light from the Moon. The light mountain is almost in some way similar to a video camera storing information using whatever format. In this special case the format relates to the way you are here looking up at these very familiar looking but difficult to look at sights. It's light and it's this very solid looking obviously real thing over there! You are seeing it somehow so some sort of light is actually arriving in your eye but the actual thing you are seeing really is this huge area of lasting and eventually growing Moon light, over there. There is some sort intensity that arrives in your eye after the shadow strikes and the light rises that is beyond description at least for me. I know it's real and is directly related to the shape effects.

It's not just the physical qualities of the Moon's light's intensity that causes such a sudden high degree of difficulty for the viewer. Another reason for the high degree of difficulty is the very basic fact that you know that the very familiar sight and then sights that you are seeing should be completely impossible and not happening at all. Very suddenly you are forced to realize what you are seeing. Looking around the valley was like looking around a valley. A very fantastic valley but it was like looking around and down at and into a place, a wide and deep valley, from above. I always think of this time as awesome peaceful seconds of wonder for me. The Object rolls with a magic gentle grace that is another sight that is describable but at the same time not completely describable. How can I put into words the way all sense of speed is lost and instead the Object seems slow and gentle as it's graceful rolling combines with a tilting and spreading effect at the same point as it seems to blend and settle into place in front of the Moon. A very tremendous wondrous sight to be sure guaranteed!

Now you can look around and that's how it is and how it feels when you look around. Looking at the valley is like looking at a valley. During the wave of the shadow it's still just like looking around. Even after the first look at the focal point of the Changed Light it is still the same, you are looking the way you would expect except you are seeing down from a very suddenly close position. Certainly this sudden look down at the back of the Awesome Good is starting to cause a great deal of the very real surprise and realization but the unexplainable intensity is still not happening at this point.

The high degree of difficulty starts and continues to climb at the instance that the Man of Light turns around to face you. That's the actual incredible starting point that is completely beyond belief. It may be that at this point when the Awesome Good turns around to his left and seems to notice you is more shock and surprise than real intense pressure from the Changed Light but at this point definitely the easy peaceful moments are over and the degree of difficulty has started to increase. This is the point were the all time ancient mystery really starts. Why does this big sight up there look the way that it does? I understand that the only way that others will be able to possibly believe that this sight exists is to actually see it for themselves. I am simply trying to describe and at the same time report and record my view of this event. I am not going around trying to do all sorts of convincing. I can see that most of my basic goals concerning the next return and some common sense preparation issues are attainable without me having to try to convince anyone of anything.

I always talk about my viewer's guide and the idea of looking. Understanding that there is a fantastic sight to be seen is a very basic starting point. In time I can see that if I have the guide out there and at least partially known it could turn out that a substantial number of people may be ready to let their eye follow the ancient series of events that will lead their eye to that ancient and very special place in the light of the Moon at the exact second that the sight of the Man of Light at the heart of the Changed Light stands up and turns around to his left and looks right back at you face to face having seemingly noticed you! Very incredible and fantastic! I know that others will know this sight and I am very fortunate to be in a position were I clearly know that I can be of help.

As the wave of the shadow travels a series of tall individual shadows emerge beyond the shadows edge. I know that this sounds strange but I am going to try to describe this series of individual shadows the way it appeared to me at the time I saw this happen. One thing I want to explain is that I did not know that I should stay looking at the tall shadows as they continued to grow. Instead I was looking to the right at the series of individual shadows and then I would look back to the left again. Then I looked back to the right and I would see the individual shadows again at a later stage in their continuing ongoing growth or lengthening and then I would look back to the left again. The result of all the looking left and right that I did was that I got glimpses of these individual shadows at various points instead of seeing them all grow longer while watching.

The leading edge of the shadow's wave starts to show some bigger separate individual shadows separating themselves outward faster and farther that the edge of the big main wave. Slightly off the center of your view and traveling away from you is a group of three now very tall shadows. It turns out that this is the first three main shadows that create forms out of the Moon's light. The central big main shadow starts to very obviously show it's greater size when compared to it's two companion shadows that are located one on either side and behind not as far into the distance of the valley.

This is it. The big main shadow! The middle shadow of the group of three shadows has to be at least twice as wide as the two outer shadows. Also the two smaller outer shadows very clearly are not as long and tall as the big main shadow and visibly trail behind the main shadow. It turns out that once your eye's gaze has made it to this point it is a natural to be able to lock and fix your gaze onto the very point of the big main shadow as it streaks towards the sight of the Awesome Good!

Suddenly like a ribbon the Black Road, the big main shadow starts to rapidly go up and down and up and down twice with your eye with it. Then the shadow's sudden left and right up and down all over the place zig zagging squiggly line starts to happen and lasts for a mere fraction of a second! Your eye had been riding the very sharp point of the big main obelisk shaped shadow and now suddenly you are looking at the very clear sight of the muscular back of the Awesome Good and then a couple short seconds later you are going face to face with the Awesome Ancient Good! Very special moments completely beyond description!

It turns out that the Object casts a shadow into the Moon's Changed Light from the surface of the Moon and now Humans live on Earth!
It's an ultimate mystery if there ever was one!

I saw it and all I have to do is go to the library and I can read about a lot of other people who also saw it. Soon it will be our turn to see and know the sight of the Changed and Lasting Moon Light sculpted and formed into the Awesome Good by the point of a very long incredible completely fantastic very dark black shadow!!!

The black darkness of the shadow's wave sweeps over the world of the Object and the shadow is good. The shadow leads straight into the Moon's Changed Light and then creates and sculpts the shape and form of the Awesome Good! The Awesome Good itself looks directly this way towards life on Earth and then this leads straight towards us humans living here on the good old Earth.

I have to guess as far as some of the bizarre seeing effects and sensations are concerned. At an earlier point on the surface of actual Object itself I looked at the leading edge of the shadow's wave back to the left against the wave's motion to the right. I suddenly found that the place that I was seeing from was very close to the Object's surface. I was seeing from very close to the ground right beside the edge of a crater. I was able to look out across the land and then actually up the slope of the next crater as if I was actually standing on the ground or hovering just above the ground actually looking into the distance across and parallel to the Object's surface. How does an effect like this work? I know how strange this sounds but I did see across the surface of the Object from a place close to the surface of the Object. Bizarre! It just happened without any sort of planning by me. It was only later that I realized that this had happened and was an unusual effect. During my view when I was seeing as I just described it was at this point that I suddenly realized a sense of scale. Even this one place between just these two craters was a very big place all by itself! For some reason I was actually thinking that it would be a long drive in pickup truck to go from were I was all the way even to the point were the land started to slope up towards the next crater. I was looking out and up at the sides of the towering rim pieces of the next crater way way in the distance across the land! Totally bizarre! A real effect!

I am mentioning this again because for one thing I have mentioned and think that at an early point while the shadow's wave is safely still crossing the surface of the actual Object itself and there is time before the shadow's wave reaches over the far upper right horizon there is time to look close down to the Object's surface. Another reason I am mentioning this effect is because it may be possible that a form of whatever effect that happens to your eye's viewing position as it travels forwards into the light riding the point of the big main shadow so that you can see down from above and behind does also seem to be at work happening all along the edge of the wave of the shadow even on the Object itself before the shadow reaches the Changed Light from the Moon. This if you don't let your eye's gaze travel with the left to right motion of the wave of the thickening black shadow I think. Instead this effect may start to work if you stop the left to right motion that carries you along and you look back left and or just simply stop and pick a point and look down at it.

Early on in the life of the shadow pick a spot and look down at it. I happened to pick a spot to the right of one of the craters. I was looking at the way the standing crater rim pieces on the right side of this crater were casting their own shadows. Actually at first I watched the shadows lengthening across and through the middle of this particular random crater and then fortunately I stopped looking further to the right once I was looking down just to the right of this crater. More and more details reveal themselves as you seem to be able to see from closer and closer and you are able to see smaller and smaller things as your eye's viewing position seems to be getting closer to the ground. This area or region was not completely blackened by the shadow yet. There were many small lengthening shadows being cast in this area at this point. As well the tops of both tall and small features were still above the height of the shadow's main wave. The amount of new details and the texture of the surface is suddenly seen in more and more detail thanks to the highlighting effect of the moving changing shadow. Get in down close and then spent a couple of precious seconds and look out across the surface from were you "are," or seem to be! You should find yourself seeing out across the land as if you were really standing or hovering there! People should perhaps consider trying out this effect.

If it's early in the life of the shadow's wave and you don't hang around looking from "down there," for to long a couple seconds just long enough to see then you shouldn't be in danger of missing the start of the big squiggly line when the shadow strikes to the right creating the muscular Man of Light. There is nothing else that is worth seeing that compares to the first instant when you see the down to the back of the Awesome Good from above and behind! Don't delay or you may very well miss the first big sight in the light! There is time to look around but there is not that much time. Things and events and sights do move right along and everything happens very fast.

I am mentioning this part of my view again because there really is an awesome effect or thing that happens along the edge of the shadow and especially at the very point of the big main central most important shadow. The place were you see from changes from were you are which is the place were you normally see from to the place you are looking at. That does not really describe the effect very well and I struggle to understand this for myself. I find it very hard to find words to accurately describe what I don't understand. I can only say that this effect is real and very mysterious and fantastic! The place were you see from races forward similar to a zoom lens effect towards the big moment and the big sight from above and behind and then the place were you see from races back! Without realizing it at the time I guess in some way I was riding in the light. That's how I can think of it so that it makes sense to me. Without my feet leaving the ground while the shadow was moving at first I saw down to the ground and then out across the land up there and then I looked around from back here on Earth were I stood and then I saw down from a place up there from just above and behind and very close to the sight of the back of the Awesome Good! A completely fantastic and real effect that I can remember, and even see! But I can't understand or explain how this very real effect happens. As a part of my viewer's guide I understand that this visual effect is of central importance because it is an integral part of the qualities of the magical Changed Light from the Moon. This basic effect is how you see from within the Changed Light. If this didn't happen then you would simply see the Changed light from the place you are standing and not from within the way seeing actually happens. Future viewer's can only benefit if they know this ahead of time instead of having to try to do all sorts of suddenly noticing and realizing as they themselves find themselves traveling forward through and into the distance of the Moon's magnificent light as they watch and follow the fantastic actions of the shadow!

The Black Road

I believe that the black road referenced to in Maya cosmology is a shadow. Not the big main wave of the thick very black shadow that crosses the crossing forward rolling Object's brilliant moon coloured surface from the left to the right turning it black but instead the black road is the big main long tall pointy black shadow that emerges centrally ahead of the other individual shadows that emerge ahead of the awesome spectacular rapidly moving wave of the shadow. The point of the big main tall long widest central very rapidly moving shadow, the black road, leads your eye up over and past the Object's upper right horizon into the distance of the Moon's incredible Changed and Lasting Light deep into the distance of the heart of the sight of the ancient Valley. Ultimately the big main Shadow strikes with it's point suddenly cutting left and right zig zagging drawing a squiggly line shaping and sculpting the incredible first sight of the back of the human form in extreme very muscular detail. This deep deep distant very special ancient place within the Changed Light where the very muscular Man of Light, The Awesome Good seen first created is the first very special sight that you see at the end of the black road.

Finding myself accidentally seeing the crossing Object while viewing the Moon using a small backyard telescope I remember and describe in detail the Shadow from it's first startling appearance suddenly forming then growing, thickening and spreading out and then suddenly dropping from under the forward rolling Object's raised area or large mound. As the Object crosses down through between the Moon and the Earth the mound emerges from over the top of the forward rolling Object towards you. The large craterless smooth curving mound evolves into a cliff like feature and then a very large overhang. The sun is located behind the viewer and the Earth. It's three nights before the full Moon early evening but it is dark. The alignment between the sun, Earth, the crossing Object and the Moon causes or results in the formation of a shadow under the newly formed overhang as the Object travels towards the Earth from the face of the Moon from the left to the right on it's slightly downwards crossing angle between the Moon and the Earth.

Sagittarius and the dark rift region of space may be the area in space where the Ancient Object appears and orbits down from. The speeding Object arrives down to it's crossing point between the Moon and the Earth from somewhere. If today's interpretations of ancient texts or depictions, glyphs etc are correct and it turns out that the ancient Maya and others are pointing at the area of space where the dark rift and Sagittarius are located then this information may be very important in understanding where the brilliantly reflective Object first becomes visible and then rolls down from.

Are there any ancient clues, myths or texts etc that suggest that any sort of astronomical body became visible or appeared in or came from this region of space? This in association with for example creation myths or ultimately in the appearance of various gods or gods etc. I do know for a fact that the appearance of the ancient Man of Light is at the root of many important things. This is the big thing that happens.

I show the Object's basic crossing flight path angle down across from the left to the right between the Moon and the Earth.

How long does it take for the speeding Object to cross down between the Moon and the Earth after it's first appearance becoming visible to the naked eye after returning from the depths of space perhaps from the dark rift region of space? The Object becomes visible appearing after being invisible, it appears and rolls down from somewhere. The Object is traveling at a tremendous speed. It crosses the distance between the

Moon and the Earth in just a few minutes and I think it will take 30 days plus or minus to travel Pluto's average orbital distance from the Sun. It will arrive home again and seem to settle into place between the Earth and the Moon and when it does suddenly the light blending and merging between the Moon and the crossing Object will begin as the Object, Adoil becomes undone.

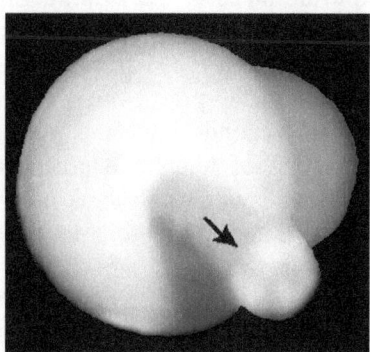

Even though the object crosses at a point well out in front of the Moon towards the Earth, it also crosses at a point that is well out from the Earth towards the Moon. I am mentioning this point because I realize that this point can help a person to better understand the image;"The Basic Flight Path."

The Moon's light has become changed and has become transformed into the sight of a fantastic valley, the actual fantastic Ancient Valley itself is seen to the right past the Object's upper right horizon with it's three steep faced mountain ranges surrounding the flat layer of white clouds that is the valley floor on three sides. One mountain range is positioned across in the distance joined at each end with the other two mountain ranges on the left and right side of the Valley. The closest area of flat misty valley floor borders on the Object's upper right horizon. You see this border area down over the Object far slope when you see from up there as if out of a remote camera the way the Changed Light cases you to see from or close to the place you are looking at as I try to describe in detail in the Return Viewer Guides.

The Object continues forward rolling down causing the giant mound to become an overhang suddenly causing the Shadow to form. The Shadow grows, creeps and stretches down before suddenly dropping leaping down to the Object's Standing Stone Crater covered surface below. Instantly the thick very black wave of the Shadow starts moving and travels towards the Object's upper right horizon through a sea of incredible massive Standing Stone Craters. Along the leading edge of the shadow a group of fence post like looking shadows cast by vertically standing crater rim pieces emerge and now proceed ahead of the Shadow's great thick black wave further faster towards the distance of the central heart of the Valley where the big main shadow ultimately strikes.

As I looked down from the place in the Lasting Light that I was being caused to see from the way the small group of individual shadows looked reminded me of fence posts for a fence around a piece of property or a house in the country or out in the prairies of western Canada. Arranged from the left to the right seconds later the individual shadows look basically exactly like a line of telephone poles stretching across the country side. You see the telephone pole looking shadows at this point from a viewing position located above this scene similar to about 300 feet to maybe around 500 feet. You have been and you will continue to see from well past the Object's upper right horizon within the Moon's Changed and Lasting Light as if out of a remote camera down at a place that is flat like land or more like flat smooth rock. As the shadows progress outwards and away from you towards the upper right they are traveling over and across the flat white clouds of the Valley floor. Instantly as they travel over and across this flat area instantly as the shadows move the layer of white clouds transforms into a flat smooth rock or stone look and spreads and builds outwards matching the progress of the group of individual shadows.

At this point as I describe over and over again in the Return Viewer's Guide that now is the time to find and follow the point of the big main shadow, the black road as it turns out. Intensely focusing and remaining looking at the point of the shadow causes the place you see from to follow and race much further past the Object's upper right horizon deeper into the distance of the Changed Light as if you were speeding down a tunnel or cave with your peripheral view focusing inwards becoming less discernible as you continue to focus on the point of the big main tallest longest widest very black shadow. Also at this point you really start to notice how your viewing position has been gradually lowering getting closer to the newly formed smooth rock valley floor.

Suddenly the now flattish looking speeding shadow goes up and down and up and down twice very quickly. Then instantly the point of the black road cuts left and right up and down drawing the squiggly line that is the sculpting and shaping and creating the muscular human form as if a statue sitting or crouching facing away towards the distance of the Valley. This shape that you are looking at is somehow made of or consists of the Moon's light after it has been intercepted and changed somehow slowed down now a solid unmoving thing a very expansive big place and at the very heart of this changed light the shadow strikes and the muscular human shape or form in exact detail is somehow caused to happen because of the shadow's very special interactions with the Moon's bizarre looking and behaving fantastic Changed Light.

I am comparing what I saw to the reference to the black road in the Maya The Popol Vuh. The black road isn't just a shadow instead it's as if the big main Shadow, the black road actually transforms into and becomes the Awesome Good, the Ancient Man of Light.

The Changed Light imprints itself into you. The ancient mystery is there. Life started on the Earth. Eventually some of that life looked up and saw the muscular form of the Ancient Man of Light. And here we are on the Earth, an exact copy of what a vertically standing crater rim piece and the Object's forces caused to happen.

A shadow is cast, the black road and we are the result. Our beginning is there to be seen. When we look all the way to the end of the black road we find ourselves. Our collective human heritage by birth starts with the truth of what happens at the end of the black road, a tall wide very long black shadow that was cast by a vertically standing crater rim piece located on the surface of the Object of the Crossing Down.

The Far Slope

I try to understand and describe various aspects and elements of the incredible scene down the Object's far sloping surface. This is the area up over the crossing Object's upper right horizon. Once you are seeing from up there the way the Changed Light causes you to see and the shadow has passed through the region of the far slope the incredible scene down over the Object's far sloping surface is what you suddenly see.

Light from the Moon and the Object and planets and stars and all light and energy sources from the background of the space that is the background for the entire Changed and Lasting Light event and is all a part of this extremely complicated event. Within the complicated

Changed Light event the incredible magical scene down the Far Slope occurs. Once the wave of the thick black shadow has passed by from the left to the right suddenly an incredible scene is clearly seen from basically directly in front of you down over the other side of the Object. Once you are looking directly at the Changed Light that is generated and or caused to happen by the crossing Object's forces you are suddenly seeing as if out of a remote camera up there as a way to compare the effect to something familiar. Look and instantly you will be seeing from the place you are looking at or from a place a lot closer to that place or thing you are looking at.

Suddenly down over the Object's upper right horizon down the the far slope it's as if suddenly you are looking at glistening sparkling fantastic place that actually looks like a city although a very special city. A place of glistening shapes that are seemingly oriented outwards towards the distance of the Valley the same way you are basically oriented.

I have always been very fascinated by the two fantastic brilliant triangle shapes that are eventually seen as a part of the incredible scene past the Object's upper right horizon down over the Object's Far Slope.

For many years I had only thought of and considered the light from stars as the possible source for the light that is seen in the form of the two brilliant triangle shapes. Once I started to read about various planetary alignments it was obvious to me that along with stars I had to consider the planets themselves as the possible source for the light that the Object's forces eventually transforms and is seen within the Changed Light. Also I have finally caught on to the idea that it's very possible, even highly likely that the sun light that is reflected from the many moons that orbit these planets may very well also play an important part in and be an important part of the lasting Light event.

I have known for a long time that there was a light source seen over the Object's upper right horizon down over the far slope that was somehow different compared to the light reflected from the tops of the vertically standing crater rim pieces that remained above the line of the thick black shadow and that this light source or double light source could be responsible for the two brilliant triangle shapes that are eventually suddenly seen.

Many different ancient texts have descriptions of various aspects and shapes seen within as a part of the Lasting light event include a wide range of colours. These colours have some sort of a source located in the background of space seen and located to the right of the Object and above the Object's upper right horizon.

Concerning the outer planets if the light from planets is intercepted captured and transformed and then stored to be eventually displayed by the Object as a part of the Lasting Light event then perhaps this could possibly help explain how various ancient peoples knew the colours of some of the solar system's most distant outer planets. The ancient viewers could have seen these coloured shapes and sights within the Changed Light and been very familiar with them and they could have wrote about them without realizing that they were from planets.

Also what other sources of light from space display different colours? The planets seem as if they display a range of colour. What about the moons that orbit the planets of our solar system. Stars themselves all look the same to me but could some stars actually display different colours? From various ancient texts that I have found to read here and there I find myself forced to conclude that at least some of the stars themselves actually do come in various colours but I know that this sounds unlikely. I just don't know enough about stars to be able to say if they do display different colours but it appears to me as though they do.

I have to make this guess just because I know what happens, the Lasting Light event, and I know where it happens, outer space. As with many of my guesses this particular guess may well eventually prove to be inaccurate. Based on the fact that I can understand some of the subjects described in some ancient texts I am still stuck with this guess because I am faced with what simple logic tells me.

Ancient people wrote about some of the different colours they observed within various stages of the lasting Light event past the point that my view of the Lasting Light ended. They saw various shapes that displayed colour. Colours that I did not see. I know that these colours had to have originated from the background of space because that's where this entire event takes place and the Moon is gray. Once again I am forced to guess that along with the planets some stars must somehow emit light that shows or displays a range of colours. Could the Object's forces cause the light from ordinary stars to be displayed in a changed way? A colourful different way compared to the way the light from these stars normally appears to us? I would have to say yes either that or stars do in fact exhibit different colours. Also there are endless types of objects and structures and energy sources and light sources spread throughout the sky. I can only assume that everything that is contained within the area of space that is the entire complete back drop or background that is seen through the Changed Light has great potential to become changed exactly the same way that the light from the Moon becomes changed. As the Changed Light Mountain rises you can see many glistening points of light within the Lasting Light layers that make up most of the smooth curving rising mountain itself. I always thought that these points of light was the light from stars however it's only been the last 15 or 20 years or so that I began to realize that there were also many other celestial objects and structures contained within the background of space behind the Changed Light event and much of it could have also been captured and changed and stored within the Lasting Light layers of the Mountain to be seen later displayed within the layers of the oval shaped plain of the third stage or phase of the Changed Light event.

I can only suppose that the ancient people would have never known the outer planets themselves for what they really were, planets but they could have been very familiar with the light from these planets because of the way they were so obliviously very familiar with the details of the Lasting light event. They were especially familiar with what was seen within the layers of the oval shaped plain of light that emerges out of the tunnel or outer sleeve of light that is originally seen as the three mountain ranges that surround the sight of the ancient valley.

The light from the tops of the vertical crater rim pieces are the source for the vast majority of the brilliant shapes seen down over the crossing Object's upper right horizon. All of these brilliant shapes are very very fantastic as future viewers will agree once they see this scene for themselves.

The two triangle shapes down the Far Slope are really something extra special even compared to the rest of the incredible bright fantastic shapes on the line of the shadow over the Object's upper right horizon down over the Object's Far Slope.

As with many aspects of this returning Object business like the light from planets and their moons as a possible source for the triangle shapes and other Lasting Light shapes and the apparent fact that it does appear to me as though somehow the stars produce or display different colours even though I know the idea or conclusion sounds completely unlikely even to me a great many things come together for me accidentally or by luck whatever you want to call it either way right or wrong I am forced to come to certain conclusions based on having seen the Changed Light.

I don't know or understand all of the aspects of this situation or all of the wide range of possibilities. At some point the Object will obscure the Moon's disc once again. The sequence of events I describe in the Return Viewer's Guide will occur again exactly as I describe. The two triangle shapes will once again occur and be seen before they are stored and carried forward to hopefully be seen again during the third stage of the Lasting Light event just like they used to be back in ancient times.

I believe a double light sources very possible Spica may be responsible for supplying the light for the two fantastic triangle shapes that are suddenly seen down the far slope.

What is the situation in regards to the stars in Orion's belt. There is some talk about how these stars match up with the three fantastic, especially the two fantastic triangle shapes located on the plain of Giza. I consider all pyramids to be models of some of the fantastic shapes that are seen within and as a part of the Lasting Light event. Obviously I have my suspicions but I simply don't know with 100% certainty exactly what light source becomes what fantastic shape that was seen and them depicted in stone by the ancient people.

I do know with 100% certainty that the shapes the ancient people admired and then copied and depicted in stone were definitely shapes that were shaped and sculpted Lasting Light. Eventually all of this will be sorted out and better understood once the Object returns.

Obviously people who understand the solar system and it's movement and mechanics against the background of space are the people that are required in this completely complicated situation. I can say what I saw and I can try to ask the right questions as I notice this or that aspect of the planets and stars locations and behaviors I am simply unable to progress much beyond this point because of my lack of knowledge and understanding concerning solar system and celestial mechanics.

Once again because of the fact that the solar system's exact position and orientation can be determined because of the Apollo Detection all of these important questions can be answered by people who work in this area. For these people determining the answers to the questions I am asking will be simple and easy once they finally realize that the situation as I describe it is what really actually happens.

I did see how the Changed light moves and some aspects of how it behaves. I saw how it can remain unchanging, then how moves and flows as a one way moving area. I saw how it can look as though the surface of the Moon with the Object lower down in front can seem to be rising and flowing upwards into space! I saw how the upward pulsating waves join in and then remain unchanging themselves the newest area of the rising Lasting Light! A truly incredible spectacular awesome sight!

The Object itself and the Lasting Light as far as I saw it are really my area of things compared to the way I an left to grasp at straws in what is the area of other people. As much as I like the way things are now with the world basically oblivious and not noticing or caring about the details of the Object or the Lasting Light event as I know it and am able to accurately describe it I know at some point that all has to change and it will change.

I will keep looking into this area of things concerning the background of space that is in place during a Return alignment. The area of space that is seen above and to the right of the Moon. The light from the area of space that ultimately becomes a part of the incredible fantastic scene that is visible down over the Object's upper right horizon down the Far Slope.

The Hand Bag Effect

Because of the fantastic cause and effect interactions between the Moon's Changed Light and the shadows that are cast into that light everything that is seen within the Changed and Lasting Light repeats and is seen again in exact detail every time the orbiting Ancient Object of the Crossing Down returns.

Everything that is seen repeats and is seen again in exact detail every time the orbiting Ancient Object returns. Everyone sees the same thing no matter when they saw or where they stood looking up. This at least as far as approximately the first 45 seconds is concerned. The part of the Moon's Changed Light event that I saw repeats in exact detail every time. It does appear clear to me that the details of what is seen repeat every time the shadows are cast into the Moon's Changed and Lasting Light. There are hundreds of incredible massive Standing Stone Craters located on the surface of the speeding forward rolling brilliant moon coloured moon size returning celestial Object that I saw in I believe probably

THE SHADOW

1972. Individual vertically standing crater rim pieces cast shadows into the Changed Light from the Moon. The interactions between the shadows and the Changed Light cause incredible results and these things that are seen at this point by the viewer are and consist of light from the Moon and the background of space that has been transformed into solid forms and shapes.

There may be the possibility that especially in the later stages of the Changed Light event to whatever degree there may possibly be some variations or an evolving predictable progression of some of the sights that are seen from one Return to the next. I am guessing that if this is the case then maybe there is a cycle that happens and once that cycle of separate Return orbits is complete everything starts over again. During the first orbit down in the beginning of every new cycle every single detail that is seen within the Changed Light would occur or be caused to happen the same way every time. If the Object does follow a repeating cycle or pattern of orbits back down through our solar system every 46years and one day I wonder how many orbits there might be in one complete cycle? One day we will answer a lot of big questions that humans haven't wondered about for a long time.

Anyways this is speculation based on a number of factors I talk about more elsewhere in the meantime the basic fact is this, everything that is seen within and as a part of the Changed Light event including the series of events on the Ancient Object's surface occur or happen exactly the same way every time. Somehow it's as simple as that. This is what happens. Without any question people will see that there are many things that are tied together by the facts of what happens when the Ancient Object crosses down between the Moon and the Earth.

All of the basic indicators are present so I have to assume that the Hand Bag Effect is just one example of many examples that show how things seen created by the shadows happen the same way from one Return to the next. This very quickly gets into areas that people need to see for themselves because they can't be imagined as being real. The shadows sculpt and shape and interact with the Moon's Changed Light. Exactly the same way the first human form is created by this effect or process by a shadow, another separate shadow created another human form later at some point this one with a hand bag looking shape that is light just exactly the same way everything that is seen is Changed and Lasting Light unless it's the hard ground of the Object itself.

The points of shadows do somehow cause the human form to be made many times. I saw twelve or thirteen examples of the human form. I am able to read about what I saw and then I realize I can read past that point and I can read about things that are seen later past what I saw. I didn't see the handbag myself. The Hand Bag Effect is thought of by some of as being an example of an ancient common theme and a mystery. Anytime I find out about an ancient common theme and a mystery I always wonder if it's related to the Changed Light. The Hand Bag Effect as I understand it at this point is very likely seen as a part of what is seen within the Changed Light event. It is clear to me that some of the many thousands of pointy shadows that are cast by the vertical crater rim pieces that neatly ring the many hundreds of Standing Stone Craters do continue to create the human form repeatedly many times and on an ongoing bases. Somehow a hand bag shape is also made as a part of the effect or process that happens.

I think that The Hand Bag Effect happens as a part of this process and this explains why it was seen and known worldwide through history. Petroglyph, Pictographs and Rock Art show this same effect with the common imagery that is seen. People looked up, and they all saw the same things. The human form as if a statue over there, up then below and in front of you made of light that is somehow looking solid like stone and smooth rock. People seeing this sight is the entire point of the Return Viewer's Guide. The Object's forces cause the emerging light from the eclipsed moon to behave in the most fantastic bizarre manor. Look up focus and stare and don't look away no matter what. The Changed Light somehow causes you to see from close up. If you look away you will have to start again. This could cause you to miss your chance to see the Ancient Man of Light being shaped and formed by the point of the big main shadow.

One of the human or humanoid looking shapes that is created by a shadow looks as if it's holding a hand bag. This is a guess. Does this effect happen more than once? Maybe however I am completely unable to say for sure I can only guess. I see this hand bag effect depicted and surrounding it I see all sorts of different scenes. This leads to many questions. One of those is the evolving effect that happens. The individual human form that holds the handbag my evolve from one look to another. I see various different characters depicted holding the same basic type of hand bag. Do many separate individuals hold a hand bag or is there one individual that holds the hand bag and that individual then transforms from one character to another? The Man of Light transforms and changes so for me it makes sense that some or all of the other human forms that are created by the shadows possibly could also transform and change as time moves forward within the changed Light.

People want to understand as much as possible about the natural world and universe, human history and our place in it. So do I and you as well probably if you are reading this. The shapes in the Lasting Light are at the root of many big things. The ancient people saw the Moon's Changed Light and it is very obvious they knew it was special and that it was also exactly what their most distant ancestors saw.

I know that it will be impossible for people to consider what I am describing here on this page. Also frequently I do have difficulty expressing myself through writing in a clearly understood manor. And here I am trying to describe the impossible sights that happen in the Moon's Light after it has been effected and somehow changed by the Object. Plus I'm trying to guess about things that seem to have happened later in the Changed Light past the point that I saw it up until.

It's as if the moon's Light has been intercepted and each wave of light separated or divided from each other, slowed down delayed and or somehow gathered up by the Object and then displayed in a smooth movie like manor matching the rate of time that is happening and moving forward within the Changed Light. Each area of Changed Light that you see eventually just changes as the time within the Changed Light starts to move forward once the big main shadow strikes and makes the human form for the first time. I have always felt that somehow when you are looking at the Changed Light you are looking at a place where time is flowing at a slower rate. Now today I understand that this might actually be the case.

The ongoing process moving forward sees hundreds of individual shadows interacting with the Changed Light and they each have their own time in the Light shaping the Changed Light. Everything that happens in the Changed Light becomes captured and stored to be seen later within the Changed Light when that particular moving as one area or wave or stored layer of Changed Light is next in turn to be displayed. As if with a movie each area of slowed down light is like one frame in the movie. One frame or one wave flows into or meshes into the next as earlier events in time are eventually seen. For example there's the human form looking like a statue over there, a solid looking thing. When he stands up and turns around to face you, you are seeing what the next one way moving area of Light looks like. The shadow does the sculpting and forming and creating that it does right on through each wave area or frame of Light. One result is a flow of separate but connected events that are unbroken. Each area or layer of light with it's own time or duration displaying in the Changed Light. At which point in succession the next frame displays and the movie continues.

Explaining the sights I saw happen when I saw the Object so that they can be understood by other people is an impossible task. I can say what happens and what things look like but for example what is Lasting Light? A thing that is outside of normal experiences and without any true known or common reference point for a person to use to understand. People may be able to see these things happen again on the night of May 26, 2018. The Object I saw orbits and I think there is a good chance it was here crossing down May 24, 1926. A 46 year and one day long orbit. We will see what happens. Sooner or later the Object will return and cross down in front of the Moon because that's what it does.

THE SHADOW

Gobekli Tepe

I'm not suggesting that I have all the answers far from it. I'm just taking the opportunity to guess and speculate and comment on some of these questions and great ancient mysteries.

4 NAMING RULES? WHAT NAMING RULES?

There Are No Naming Rules!

One of the most ancient human traditions that I can think of or that I know of is to name the Object! It's the human thing to do.

That's what all of or most ancient ancestors did. They all named the incredibly massive speeding moon sized very reflective brilliantly shinning moon coloured forward rolling standing stone crater covered Object that rolls down across between the Moon and the Earth.

Either that or their most ancient ancestors had already named the Object and those various different names were handed down through time to them and they were following their traditions and they were using the ancient names they had always known. Everyone's ancestors named the Object and or used ancient traditional names. One of our most ancient human traditions is to name the Object.

Another very very ancient human tradition is to name the many various elements and human statue like sights that are seen within the Moon's Changed and Lasting Light that are themselves Lasting Light. This basic observation that will be shown to be a basic true fact I believe is at the root of an area of study and discussion that has no bounds. Within this area is contained the entire human condition on Earth from the very very best to the very worst that humans have to offer towards each other and the environment around them. The many various elements and statue like sights that are seen within the Moon's Changed and Lasting Light are profoundly important. Unfortunately the basic loss of knowledge and understanding regarding what or ancestors were actually looking at and then doing when they named the things they saw has developed into a religion based situation that is seemingly nearly entirely disconnected from the facts and the knowledge of what happens at the root of it all. This continues to have many extreme and negative effects on our lives today and the prospects for the future. I hope that once the people of the Earth see the Object and the Changed Light again a positive shift or change in human thinking towards peace and cooperation follows soon after.

The below text is a favorite of mine. As with most ancient Return descriptions I have no idea how this text is thought of today or what people think it is or what it means or what it is about. I do know that the ancient people wrote about the Returns that they saw. The sight of the forward rolling Object crossing down between the Earth and Moon and the sight of the Moon's drastically Changed Light were the literal sights that the ancient viewers wrote about. If a Return is the big sight that is being described by the writer ancient or very ancient or modern then that's what it is about without any question. The ancient names and terminologies differed but the Object's return and the repeating sights that are seen in the Changed Light has always been the focus. The sight of the muscular Man of Light or Awesome Good as I describe in detail in many places is the main reason why I know that this is so. If you know what happens during the Object's crossing down in front of the Moon and you are able to recognize the repeating sight's and the repeating sequence of events that the ancient writer describes then it's clear and obvious and unavoidable because everyone sees the same sights and the same things happen as a part of the Changed and Lasting Light event. This is a basic fact.

I have no idea if anyone out there takes the below text and some other ancient Return descriptions seriously or not. I do and I know that I can learn about what happens after my view was over by reading what was written long ago. It doesn't mean I know it the same way like I would know it from seeing it but sometimes I can get a sense or basic general idea about the things that I didn't see because either they happened after I wasn't looking any more or they didn't actually happen at all during the 1972 Return. I don't believe the oval shaped area of Lasting Light happened. A question I have in this regard is will the oval shaped area of Lasting Light happen again perhaps the next time the Object returns? Since it stopped does this suggest it has stopped happening for all time or as part of some sort of ongoing cycle will we see the oval shaped area actually happen again? I hope so because if it does that would mean that we will see the most spectacular part of this event the way it occurred in the most bizarre ancient descriptions we have access to today.

The ancient writers were trying to describe the sight of what should be impossible from the beginning when the Object first becomes visible through to later descriptions of the Moon's changed Lasting Light evolving and somehow in action. This is a big reason why these texts sound so unbelievable when you try to read through them. Somehow I saw the Object and the early stages of the Lasting Light event. I saw those early impossible sights. I know that these ancient Return descriptions describe real sights and a real event. I know that the below text along with some other ancient Return descriptions that I have read, need to be taken very seriously not just by me but by everyone.

For everyone out there you can only wait and see. For me I know for a fact the some of the ancient people refereed to the Object that I saw as Adoil. Because of the spectacular descriptions of the ancient Object returning in Enoch writings, Adoil is the name I have decided to borrow and use for the Object.

More descriptions of Adoil and some of the Lasting Light sights

(Enoch 2)
Chapter 25, XXV
1 I commanded in the very lowest, that visible things should come down from invisible, and Adoil came down very great, and I beheld him, and lo! He had a belly of great light.
2 And I said to him: Become undone, Adoil, and let the visible out of you.
3 And he came undone, and a great light came out. And I ... in the midst of the great light, and as there is born light from light, there came forth a great age, and showed all creation, which I had thought to create.
4 And I saw that (it was) good.

Adoil, the descending Object set to roll down across between the Earth and the Moon. The Object is not visible at first. Once the Object gets close enough to the Earth and our sun sunlight from our sun reflects off the Object's very reflective crease marked bulging indentation lined shinning Standing Stone crater covered surface. At some point the speeding Object becomes visible as it approaches it's crossing point between the Earth and Moon.

It was early evening just after dark when I first saw the Object and at this point it was completely covering the Moon. To a viewer who knew were to look in the sky for the returning Object well before it crosses in front of the Moon, it might have appeared as if a new star was suddenly visible that night. Would the Object be bright enough to be seen before dark earlier in the evening? Certainly the size of the Object's point of light would appear to grow drastically larger very quickly at least past whatever point as the Object neared it's crossing point between the Moon and Earth. The forward rolling Object although probably actually at least slightly smaller than the Moon it is actually very huge and full and and massive. A very bright reflective very impressive large fantastic crater covered astronomical object. Apparently the Object can appear to be a star that falls down. I can relate to that although the speeding Object gracefully rolls past across in front of the Moon on it's way down below the Earth towards the sun as opposed to the Object, a star just simply falling down the way a stone falls when you drop it.

The forward rolling Object becomes, "undone." The Object obscures the Moon. Suddenly the Object tilts towards you and at the same time appears flattish looking instead of continuing to roll towards the viewer the way it had been previously. At this instant you are also seeing the Object from very close up. Your viewing location or the place in the suddenly Changed Light where you are now seeing from is positioned above the surface of the Object that precedes the mound or raised area that is about to come into view from over the forward rolling Object. The light from the top of the raised surface feature somehow remains in place and the Object seems to open wide spreading down and at the same instant the sight of the magnificent ancient valley of old is visible to the right of the Object and over the upper right horizon of the Object. The light from the "magnificent desolation," of the Moon becomes the sight of the Magnificent Ancient Valley. The Object has become,"undone." That describes what actually happens. Ever since I found and read this particular ancient description of the Object's return to it's crossing point between the Earth and Moon in Enoch 2 I have come to think of the Object as becoming undone when it tilts instead of rolls.

The Moon's Changed Light, the sight of the Valley becomes the amazing magical rising Light Mountain instantly after the light's inward pulsing of the short transition period ends. Light from Light complete with the incredible sight of the very muscular Man of Light standing up and turning around to go face to face with you. The Awesome Good, the Man of Light is the very heart of the entire Lasting Light event.

Then the third phase or stage happens next. I saw the Valley and the mountain but my view ended at that point and I have to guess about what happens next or rather what used to happen next. Once again "Light from Light." The oval shaped plain of light is transported towards the Earth in layers down within the Light Mountain "Light from Light." I don't know the full scope of this area of understanding but I do know that this area of understanding exists. The oval shaped plain can emerge out of the sleeve of the mountain. Descriptions in ancient texts describe the oval shaped plain and certainly this used to be the most spectacular part of the entire Lasting Light event.

It's always about the sight of the same object and the same spectacular celestial event. The names depend on who's seeing the Return and who's naming or who has already named the sights that are seen and who's using those names that are already chosen by others. Not all obviously but many ancient texts are about the same thing, the returning Object and the Moon's awesome magical Lasting Light. The names people chose to use or the names people decide to use or invent describe the exact same sights.

Everyone has the right. There are no rules as far as I am concerned, at least rules that have to be followed. If you want to follow some sort of rules or customs concerning the names for the various aspects and elements of a Return and the Lasting Light then you should be free to do so. Also again you should be just as free to decide not to follow any customs or rules.

I know that there are more ways of thinking about these sorts of rules and customs than I could ever imagine. In my own opinion I think that more diversity is better. People, humans are diverse and humans have rights or at least they are suppose to have rights and they should have rights. Rights that are equal! It's all about the same thing, a Return. People should be free to think about this any way that they want to think about it. Everyone's opinions and traditions are equally valid. This is my own opinion and I realize that other people some other people may have opinions that may be even strongly opposed to my own opinions in this area. It turns out that's how it goes with this entire returning Object and Lasting Light event business. Not just for me but for most everyone it seems at least up to whatever degree. I know that opinions already do

NAMING RULES? WHAT NAMING RULES?

differ and will continue to differ and I think that's a big part of the point.

Soon with the shock of the facts of the Return event we will be in a suddenly different and difficult situation. I can only suppose that as usual there will still be lots of very different opinions especially concerning all the difficult new or rather difficult old facts concerning a Return event and especially what happens and what is seen within the Changed Light.

I think that we can all still get along with each other no matter what. In my mind it seems as though getting along with each other should be easier after seeing the returning Object and especially after seeing the Moon's Lasting Light. I hope that seeing a Return, another safe Return will help people everywhere.

In my humble opinion in this difficult area there is no wrong name for the Object or this or that element that is Lasting Light. How could anyone person or group have the authority or power to determine what others think or what names are used to describe the various elements of a Return or what it all means? Of course the opinions of every single person and every single group of people are equally valid and couldn't possibly be nullified. This is already a case were everyone is already right at least basically as far as I know.

After the Object's next return people will be comparing the texts they've grown up with and know with the sights they've just seen in the Moon's Lasting Light. A tremendous time in our lives and our Earth's history is about to occur! There is a very long list of correct names that are related to the Object and elements seen in the Lasting Light event that it generates. In our modern day not every name is matched up with the literal element or sight that it was originally intended to describe. The ancient people knew what the various terms and names they used, described but somehow through time a great deal of knowledge and understanding has been lost.

Knowledge has been lost but many times the ancient names themselves still seem to be the actual ancient names even though some of the meanings have been lost. Very soon a great deal of knowledge will be recovered and the origins and original meanings of many names and terms will become obvious. Also it appears to me as if many times known names for the Object itself are still associated with the correct subject, the Object. It's a very long list. Many people even know this but they just haven't actually seen the Object itself but they are able to easily understand the basic idea and they know that the name they use or refer to describes some sort of large celestial object.

It is my hope that the sight of the Object and especially the sight of the Lasting Light will bring people closer together compared to the situation around the world today. I believe that if people really stop to think about this this entire event the incredible human shapes and forms seen within the Moon's Changed and Lasting Light and human life they may start to see that a great many of the differences between people are actually imagined and are not real. I think that it is true that there are many differences between people and I think that in part this is what makes humans unique and special. The fact is, even humans that are the same, are different from one another. In this respect we are all the same. All throughout most of my life I have been fortunate to know that we are all the same, different but the same. All of our ancient ancestors saw the Object and knew what the Moon's Changed and Lasting Light looks like and what is seen within and in this respect again all people are equal and the same.

I sincerely hope that after we have had our turn to see and get noticed by the Awesome Good going face to face with the face a muscular human man of solid lasting Moon light who was certainly once thought of as being a god or even God himself, the ancient human tradition of respecting having automatic empathy and caring for others will once again be one of our most important customs.

I saw the Ancient Object from Mississauga Ontario Canada. That's were I was standing, when I looked up and saw the Object of the Crossing Down and the Changed and Lasting Light from the Moon! That's were I was standing when I got noticed and went face to face with the incomparable ancient Man of Light! My Mississauga Sighting.

The ancient history and the ancient traditions and the ancient stories that were handed down through time from the ancient ancestors of the people of the Mississaugas of the New Credit First Nation to our modern day are real. I was standing in the ancient homeland of a great people. I am very fortunate to have seen the Ancient Object's last return orbit down across in front of the Moon. I feel very honored to have been standing in such a special ancient land just less than a mile west of the Credit River and just less than one half mile north of Lake Ontario.

NAMING RULES? WHAT NAMING RULES?

The Awesome Good, the Ancient Man of Light looked up directly down at me directly face to face were I stood in the ancient land of The Mississaugas of the New Credit First Nation. Everyone who has ever stood in the ancient land of The Mississaugas of the New Credit First Nation and looked up and saw the incredible sight of the Awesome Good looking down directly at them were they stood felt special. I am able to understand that people in that very same land maybe even standing close to were I stood looked up and saw the sight of the incredible Awesome Good, the Ancient man of Light.

It was the same for everyone worldwide no matter were they stood or when they looked up. If people looked up at the right time they saw and they went face to face with the ancient Man of Light, they got noticed and they felt special. It turns out that the sight of the incredible Awesome Good belongs to everyone and everyone is special!

I saw the Ancient Object and the Moon's Changed and Lasting Light from the ancient land of the Mississaugas of the New Credit First Nation

Anishinabe Nation

Anishinaabeg translates to

"People from whence lowered," "The good humans,"
"Beings made out of nothing," "Spontaneous beings"
since the Anishinaabeg were created by divine breath

Mii maanda niigaan da-ni-giizhgak ge-mno-aabjitooyan.

Mii maanda sa ge-waabndameg giinwaa waa-bi-zhaayeg.

Ezhi-gchi-nikaayan ni-izhaan.

In days to come you will have good use of this.

Such are the things you will see, you who will come here.

Go to your right.

5 "COUNTENANCE"
IT'S SHOOCKING BUT THIS IS WHAT HAPPENS

I'm worried, that's nothing new however every now and then I see something or I find out something that reminds me that I can't even begin to possibly imagine the size of the shock to come. One thing is certain and that is that once the facts of what actually happens between the Earth and the Moon become known again every person will be effected to some degree in whatever way sooner or later. As always humans will continue to go forward all men and women created equal in the image of the Awesome Good. People already know this in our hearts thanks to our ancestors who intimately knew about the The Object of the Crossing Down and the fantastic Lasting Lights shapes and human forms that are seen created by the shadows that are cast by the crossing Object into that Changed Light reaching back in time to our very origin.

We are an exact conscious living matter copy of what the heart of the Changed and Lasting Light looks like and and how it is shaped. Sitting or crouching facing away towards the distance of Changed Light from the Moon which is looking like a valley at this point, you see an extremely precisely detailed muscular human shape that is solid Lasting Light from the Moon as if a statue. The big main shadow that I describe in detail in the Return Viewer's Guide caused and created this shape. You very suddenly see this shape in front of you when the shadow's point strikes perhaps the surface of the Moon, from a viewing point or position located just above and behind this shape within the Changed Light as if you were standing on a ladder looking down to the ground a small number of yards away. Very clearly visible you suddenly find yourself looking down in front of yourself and you see the back and the right side of a crouching or sitting very muscular man.

I started this page during the time that Pope Francis was visiting the United States. I watched cnn coverage and also the cbc and ctv coverage. I was struck and impressed by the devotion and passion and the love that the public showed at all the various events the Pope attended. At various points prayers were lead by the Pope and were shown live on tv. Over the years I have heard the word "countenance," used and once again during prayers I was hearing this word countenance used. It occurred to me that I should look it up and see what it's suppose to mean or be about. This particularly so because I was once again hearing countenance used in association with a reference to "his face." Incredibly after all that you have seen happen once the Moon's Changed Light is first seen suddenly you find yourself seeing face to face with the face of a human shape that is solid Lasting Light. Once this intense staring contest starts it continues and is central to and is at the heart the entire Changed and Lasting Light event.

"Countenance
1. Appearance, especially the expression of the face:
2. The face or facial features.

Countenance, a reference to the literal or metaphorical "face of God"

Countenance' refers to the human physical appearance."

Once again, I'm worried and again that's nothing new however every now and then I see something or I find out about something that reminds me that I can't even begin to possibly imagine the size of the shock to come. This reminds me of when I first heard the definition of the word resurrection. The regular Greek word for resurrection appearing throughout the New Testament is anastasis. This is a compound word comprised of ana, which means "up," and stasis, which means "to rise," or "to stand." Thus, anastasis means "to rise up" or "to stand up."

The Awesome Good is how I think of the Man of Light who is at heart and center of the Changed Light and Lasting light event. The muscular Awesome Good does both, he stands up and turns around and then gently rises up twice on an angle upwards oriented face to face towards the place where you are seeing down from within the Changed Light at this point within the flat layer of the white clouds of the Valley floor.

This human shaped form of solid Lasting Moon Light is the heart and is the focal point within the Moon's Changed Light and was created and formed and shaped into the precisely detailed exact form of an impressive muscular man by the point of a shadow that was cast by a vertically standing crater rim piece located on the surface of the crossing Object I describe.

This incredible human shaped Changed Light form changes and evolves and moves and continues to exist or last. Once the flat area is seen to his left suddenly from his original crouching or sitting position facing directly away from you in one fast motion this muscular male human form turns around to the left, his left and stands up to face you directly, the viewer. It's still a shape that is the Changed light from the Moon that was then further shaped by the same big main shadow that led your eye to this spot within the Changed and Lasting Light originally. Face to face you are now seeing something that happened earlier as the shadow continued it's ongoing interactions with the Changed light. A record of sorts is stored within the Changed Light's many waves or layers of slowed down light then to be displayed or seen later. At a basic level and basically overall that's what happens. Once the shadow drops shadows cast from the crossing Object interact with the suddenly drastically Changed Light from the Moon and then incredible things are seen because of the cause and effect interactions between the shadows and the Changed Light from the Moon.

Finding out that resurrection means to stand up was shocking for me. Since I was twelve years old I've known about how he stands up and turns around and rises and you go face to face at this point and beyond into his new second look or appearance. Finding texts that match with what I saw during the Object's return happens however now and then it's just more shocking than other times. For me this is now very true with countenance as well and is the reason why I'm writing this page. This really reminded me about the deep and far reaching implications that I can't imagine and the size of the shock and surprise that people will be suddenly facing also something I can't imagine.

I have always had to try to be ready for whatever the next shock to do with the Object was going to be about but I am never completely really ready. A basic question I have to ask when I see something or read about something that is about or might be about the Object and the shapes and things seen within the Moon's Changed Light is this; compared to what I saw for example what else could the ancient text or ancient myth or common ancient theme etc be describing? Often it's very clear. After you see the Object and the Changed Light the same effect will happen to you where you will just know right away for example that an ancient text you are reading or know matches exactly with something you just saw in the Moon's Changed and Lasting Light.

You will ask yourself the question, compared to what a person sees in the Changed Light for example what else could the ancient text be describing? If it's clearly obvious that the text is about the Changed Light or the crossing Object itself you will understand that fact instantly. There are many things and topics recorded and described in ancient texts. One of those topics is the celestial Object that crosses down through between the Moon and the Earth and the Changed and Lasting Light event that the Object causes to happen to the light from the Moon and the light from the background of space. This is a basic fact that will be suddenly shockingly apparent to everyone.

It is my belief that among those people who lead us the leaders that have moral authority and the true love and respect of those they lead and others as well will not only survive the truth of the Changed Light but they will be propelled into the future by knowing about the Object and especially by knowing about seeing the Moon's Changed and Lasting Light.

The ancient mystery is there, a shape like a statue is caused to happen when at shadow cast by a vertically standing crater rim piece on the surface of the speeding forward rolling Object interacts with the Changed Light. This shape is solid lasting Changed Light from the Moon and is in the form of a muscular man.

The Changed Light's magical imprinting effect starts instantly as soon as you are looking at the Changed Light. The way you find yourself suddenly seeing from close up also happens instantly as soon as you start to see the Changed Light. Directly after you are finished seeing the Changed Light the after glow effect that you'll see is stunning and pronounced. The outer edges of this shape you will be seeing small and just out in front of you matches the entire overall outer shape of the Changed Light from it's first instant onwards when it was initially the sight of the crossing Object and the Ancient Lasting Light Valley to the right. The same effect that caused you to see the immediate after glow of seeing the Changed Light, an effect I describe as the Large Shape Effect, the same overall shape of the Changed Light when it's in the form of the Valley is the same cause or effect that somehow allows you to be able to actually see that exact same glistening cloudy effect again even many years later.

I think that this same after effect is also directly related to the answer to the question; why are we an exact copy of a shape that is light? The Changed Light has special qualities that I believe lead straight to the heart of the ancient mystery and The Shape Effects is how I think of this effect and my best attempt to try to begin to describe it.

Either what is seen in the Changed Light and the Man of Light are connected to us humans down here on Earth or there is no connection. For me it's clear and my best guess is that some life eventually looked up and eventually some of the life that looked up and saw the heart of the Changed Light was caused to eventually somehow physically copy and become the shape that it was seeing. We are an exact copy, we bear the image of an primordial statue like shape that is Moon light that had earlier been changed and transformed by the crossing Object, then next to be sculpted formed and shaped into the solid 3 dimensional shape of a muscular man by the point of a shadow that was cast by a massive towering vertically standing crater rim piece located on the surface of the crossing Object. The basic bottom line is a fantastic thing happened and it still happens! You can see the exact same sights that have always been seen the next time the Object rolls down across between the Earth and the Moon.

The Ancient Object returns and crosses down eclipsing the Moon. When the Moon's light emerges from behind the speeding forward rolling Object it is seen to the right and in the area above and to the upper right relative to the Object's upper right horizon. Somehow the light from the Moon has become drastically changed in appearance and behavior by the speeding crossing Object's forces and influences and actions.

The Object's raised surface feature comes into view up from behind the Object up towards you from over the top of the forward rolling Object. This giant mound becomes steep curving cliff then a massive overhang. A large shadow is suddenly caused to form under the mound clinging to it's ceiling as a result of the Object's continuous forward rolling motion. The growing shadow spreads out left and right and thickens and then it

"COUNTENANCE " IT'S SHOOCKING BUT THIS IS WHAT HAPPENS

reaches and stretches down for an instant before it suddenly drops down to the Object's surface below. Instantly the shadow starts moving across the Object's surface to the viewer's right on an angle towards the upper right region traveling through a sea of massive towering Standing Stone Craters. Ahead of the thickening very black shadow's moving wave long narrow shadows that were each cast from individual vertically standing crater rim pieces are seen.

Next a group of three main shadows emerge ahead of the rest of the individual tall shadows that lead the shadow's main thick black wave. Looking at the point of the larger central main shadow of the group of three shadows causes or results in the viewer seeing that this shadow's point suddenly sculpts and shapes and create the heart of the Changed and Lasting Light and that is the sight of the back and the right side of the ancient muscular Man of Light. Then the things you see and everything that happens progress from there with the way the shadow continues to progress sculpting and shaping through the next areas of moving Lasting Light causing what the viewer sees next to flow or further progress.

Next after the Man of Light drifts upwards twice while the viewer is still looking down from the same original unmoving viewing position from above and behind, suddenly next the viewer sees what is seen within the next area or wave of moving Lasting Light. The shadow's time earlier sculpting and shaping further in the Changed Light happens and is seen like a natural movie or action or movement and the next thing that is seen is the Man of Light turning to his left and standing up all in one fast motion to then be oriented directly towards the viewer face to face. It's Light that has been shaped by Shadow into the shape of a human face with an upturned stare that results in sudden startling direct eye to eye contact for the viewer. A startling direct dramatic intense face to face situation.

Next this shape continues to drift upwards towards your viewing position, the same place were you first saw his muscular back from. At this point you are unmoving still seeing from the same spot where your earlier forward motion stopped. At this point you are also still seeing from down below the top of the layer of the flat white clouds of the valley floor. You are still seeing from within the white misty indiscernible cloud effect now going face to face as this next area of formed and shaped Lasting Light gently drifts upwards on an angle towards you twice. I used to always think of this as the big noticing. The very real feeling that I had was that I was being looked at and scrutinized ever harder as the face of the Awesome Good got nearer rising closer. Face to face is what is really seen and what really happens. And it keeps happening locked in with incredible increasing indescribable intensity and difficulty as I try my best to describe in the Return Viewer's Guides.

As impossible as this sounds unless you've seen and know this for yourself and I know I could be mistaken although I don't think so, I believe that the face of the heart of the Changed and Lasting Light, the face of the ancient Man of Light is seen carved in stone and is the face of the the Great Sphinx of the Giza Plateau. I believe this is Man of Light's third evolution or look after his initial muscular look and his second look, the rising Lion Man. Properly seen from a place that would be up in the air above and in front of the face of the Sphinx where you could look down on an angle that matches the angle that you are looking down on when you first see him sitting or crouching facing away towards the valley. Then this is when when he first gets up and turns around to his left all in one motion to then go face to face with you. This is preceding and leading up to the next instant in time, the start of the first initial face to face seconds that you see before the entire area of Changed Light rises. Before the Changed Light's repeating inward pulsing happens as you look down face to face you are looking directly into the upturned eyes of the nearing face of the Man of Light he gently rises towards you twice upwards along this same angle that he is facing you on. Look at the angle that the Sphinx is looking up into the sky and that's the same angle you find yourself looking down on him when you are going face to face with him.

An extremely important fact to remember is that the sights I am describing were caused to happen within the Moon's Changed Light because crossing Object's forces changes the way the Moon's light looks and behaves. Then the vertically standing crater rim pieces on the crossing Object's surface cast individual shadows in the direction of and into the Moon's Changed light. An interaction takes place between the shadows and the Changed Light. Suddenly the light lasts and remains solid in place. Everything seen and everything that happens between the shadows and the Changed Light has to happen in a repeating exact way for me to have been able to see these sights at all. I saw the muscular Man of Light in perfect detail. I can read about what I saw. The only way this could possibly happen the way it does is either with the whole solar system is involved or at minimum, the crossing Object, the Moon, Earth and sun are in such extreme synchronization that when the pointy tips of the shadows strike and start to sculpt and shape and create in the Changed and Lasting Light they do so in such a way as to sculpt an incredible statue like figure of a perfect muscular man. There must be some sort of orbital cycle in place that is in perfect balance and timing if the shadows can do this in such precise detail. It's all to big and beyond me to understand but I can try to ask some of the right questions. I have always been in total awe and wonder and respect when I think of the Man of Light, The Awesome Good.

The way the Man of Light looks changes. I saw the first two looks or versions that happen. I believe there may be three types of looks that happen namely the one third man two thirds beast and because his face is a match with the face of the Sphinx I believe that this is the third evolution of the Man of Light . It's very possible there may be more types and variations as well I don't know. The first version is the muscular man sitting then standing up to his left and turning around face to face. Then face to face with you he then closes the distance rising twice directly towards your position along an angle that matches his upwards stare. The second version happens when he stops after the second time he moves upwards. Suddenly the light's rapid repeating inward pulsing happens. Since I was a boy I have always thought of his second transformed appearance as the Lion Man at this point and moving forward from this point in the Changed Light event. The instant you see this new sight, this new lion man shape or form that is Changed and Lasting Light, everything that you see very suddenly rises in extremely dramatic fashion. This is the time of the rising Light Mountain and presents the most difficulty for the viewer as I try to describe in the Return viewer Guides. Face to face seconds that always happen and can be seen by anyone who looks and doesn't look away. Uninterrupted direct viewing is key to seeing in the Changed and Lasting Light.

Within the Changed Light you will suddenly find yourself seeing from the place you were just looking at. A hard to describe viewing effect but basically that's how it works. If you look away you will have to start over. Look away and you might not get to see the big main shadow's point's sudden squiggly zig zagging motion as it creates the muscular ancient Man of Light out of and with the solid looking Changed light from the Moon. Look away and you might not get to see and go face to face with the Man of Light. Please follow the viewing information in the Return Viewer's Guides and whatever you do once you are seeing the Changed Light, DON'T LOOK AWAY!

This is much more than basic viewing advice.

If you are seeing the rising Light Mountain Phase happening you will be shocked and you will suddenly find that you are not only still seeing from up close up there, you are physically shocked because instantly when the light starts rising you become aware that you can feel an incredible indescribable pressure that flows into you as you see the Changed Light rise. Another huge reason for even more shocking surprise is that you are now also seeing more human forms arrayed around and beside and around him in two curving semi circle arches that match and face towards the curve of the upper right horizon. Somehow the long tall shadows that the massive towering rim pieces cast can cause the human form to be the result of the interactions of the shadows and the Moon's Changed Light not just once but instead many times. Somehow this is what the shadows cause to happen.

Countenance
1. Appearance, especially the expression of the face:
2. The face or facial features.
Countenance, a reference to the literal or metaphorical "face of God"
Countenance' refers to the human physical appearance.

Numbers 6:25-26 King James Version (KJV)
25 The Lord make his face shine upon thee,
26 The Lord lift up his countenance upon thee,

Numbers 6:26

International Standard Version
May the LORD turn to face you,

Douay-Rheims Bible
The Lord turn his countenance to thee,

World English Bible
Yahweh lift up his face toward you,

English Standard Version
the LORD lift up his countenance upon you

GOD'S WORD® Translation
The LORD will look on you with favor

"COUNTENANCE " IT'S SHOOCKING BUT THIS IS WHAT HAPPENS

The facts of the ancient crossing Object and the Moon's Changed Light are shocking for me to know but I'm stuck with them because of what was a complete lucky fluke while taking my turn using a small backyard telescope. I saw the ancient crossing Object rolling down and I saw the Changed Light event that it causes to happen with the Moon's light and the light from the background of space otherwise I wouldn't know anything about this just like other people. Clearly obviously this is not an easy thing. I have my opinions as to what some of it might mean. Soon everyone will know and be confronted by the facts of the Object's return and what happens and what is seen in the Lasting light. Opinions will be formed as to what it all means and what's it's all about as in keeping with our human traditions and our ancient human heritage.

Hopefully the truth of the Changed Light will cause us humans to agree that we need each other and we have to get along and cooperate in order to save our planet and ourselves. I believe that soon we will realize, we will be caused to realize that this is the single most important task before us all.

6 THE BIG FULL CIRCLE

I was watching TV show about the Apollo Moon landings a couple years ago back in 04. While I was watching the show suddenly a thought struck me. As usual whenever a thought strikes me to do with this business I am always wondering why I didn't think of it before. Anyways that's sort of how things go for me sometimes.

There I was watching Neil and Buzz walking on the Moon when I was suddenly struck by the totally obvious. This is a full circle thing, a big full circle thing! There I was watching not just one man but two men walking on the surface of the Moon, incredible!

I really hope that one day as many of the men who actually walked on the Moon as possible will truly know this. I think that it turns out Neil Armstrong couldn't be more right about the small last step that he made down onto the Moon's surface and the "Giant Leap For Mankind," words that he spoke.

When Neil took that final step down off the ladder he was literally completing a circle that must of started at the very very beginning of the evolution of the solar system and the formation of the Earth, the Crossing Object and the Moon system. This very big circle started at the very beginning all the way back to the time when our star ignited and began to produce light.

What a fantastic thing!
I feel very fortunate to actually know this! Way to go Neil and Buzz!!!

THE BIG FULL CIRCLE

At some point vertically standing crater rim pieces were formed or produced around the craters of the Object that rolls down across between the Earth and Moon. At a critically precise point a certain particular most important massive towering standing crater rim piece casts a shadow that then passes over the Object's upper right horizon region into the Changed Light of the Moon.

This particular main most important shadow is wider and taller longer than the two similarly shaped shadows that also travel along with it close alongside one on either side. The main shadow is also bigger than all of the rest of the tall individual shadows to the left and to the right that it travels with ahead of towards and into the Moon's Changed Light that is already seen to the upper right at this point, the sight of the Valley.

The shadows continue up to and past the Object's upper right horizon into the Moon's Changed Light that is seen above the Object's upper right horizon and to the extremely deep distance straight ahead and to the extreme right. The Moon's very changed looking and behaving light at this early point in the Changed Light event now looks like and is in fact the tremendous expansive much written about incredible Ancient Valley of old as it has always been seen back through time to or origin. At least it appears to me as though it has to be this way because how could this incredibly bizarre extremely complicated precise improbable celestial event be happening now if this was something that didn't start happening from the very beginning of the existence of the Object and the Earth Moon system?

This is something you can see happen if you are looking at the Moon when the Object rolls down the next time it arrives back home to cross through between the Earth and the Moon. That's the basic idea and what my Return Viewer's Guides are about. When I describe what I saw I am also describing what has always been seen and what will be seen by anyone and everyone who looks up and sees a Return event past, present or future. All the of the Object's actions and the resultant sights that are seen repeat in exact detail because everything that the Object does happens and occurs and repeats exactly the same way every time it returns. It really is as basic and incredible as that.

It could turn out that a single individual crater's standing rim piece's may ultimately be shown to cast the shadows that make up the first main group of shadows including the three main central most important shadows that do the first sculpting and shaping in the Moon's Changed

Light. I try to detail my thoughts and guesses concerning this crater's possible location just up over and past the mound or overhang's position on the Object's surface beyond. A single individual vertically standing crater rim piece does cast the central most important shadow. One reason why I think that the most important crater rim pieces are located beyond the mound raised surface area is because that area is positioned higher compared to the Object's surface area where the original big thick black shadow first occurred and then traveled towards the Object's upper right horizon. As the Object continues rolling forward the craters beyond the mound become positioned in the sun light and they are located at the highest point and this matches with the way the main group of individual shadows appear and become visible further past the big wave of the shadow. As the shadow moves across the valley's floor magically transforming it's flat white layer of clouds into a surface that looks like flat smooth rock the important individual shadows emerge from along the shadow's leading edge. I believe that this is the point when a crater located on the top of the Object is responsible for casting theses shadows

Suddenly as you are watching to the right following the point of the big main shadow. A a special place in the distance of the Valley the now very tall long and strait shadow's point goes up and down and up and down twice. Then instantly it starts to draw a squiggly zig zagging line in the very special incredible Changed Light from the Moon. The Moon's light has been made to last as if a solid thing. The shadow interacts with the Changed Light and solid forms and shapes are created. Somehow incredibly a muscular human form is suddenly there and at the center or heart of this entire event at this point. And then the Light moves and changes and then rises! Incredible but these are the repeating sights I am always struggling to try to describe. Everything that you see happen does so in an exact repeating fashion every time the Object returns and it's actions causes this celestial occur. These are the actual ancient sights that happen and you can see them too.

Somehow incredibility and very suddenly there he is, the first sight of the back of the fantastic ancient and very muscular Man of Light created instantly by the interactions between the continuously moving shadow's point and the Moon's Changed Light. You are looking at a form that is light as if a statue over there. The Object's powerful forces change the way the light from the Moon looks and behaves. Massive vertically standing crater rim pieces cast shadows from the crossing Object into the Moon's now suddenly Changed Light.

There are many thousands of individual shadows cast from the Object's surface crater rim pieces into the Moon's Changed Light and at this point that light is the form of the previously created sight of the Valley. The main shadow travels away from the viewer to the upper right into the sight of the Valley and then it strikes perhaps on the Moon's surface, and this is the point when the sculpting and shaping action by the shadow in the Changed Light is happening. Instantly right in front and below the the place in the Light where you suddenly automatically find yourself seeing down from you see the first form that is suddenly there and it is the sight of the back of a muscular man as I try to describe in many places.

The shape you find yourself looking at is not a living being the way we naturally normally think about it. This is a real problem for me when I try to convey the basic idea of what I am trying to describe here. People usually are suddenly thinking the only thing that they can possibly be thinking and that is that basically no matter how I try to describe it they imagine that I must be simply nuts and that I am imagining that I saw some guy looking at me when he few by on his own planet. People can't help it what else can they possibly think about me and my story? lol That is just the way it is and of course I completely understand. Somehow the shadow's actions cause an area of the Moon's Changed Light to become sculpted into the shape of a muscular man. This point of fact is very hard to successfully convey to people or more like impossible to convey to people .

A basic point to remember is that everything that isn't the Object itself is Changed Light no matter what you find yourself looking at and seeing. This basic fact is not so obvious for a viewer. And even more difficult for a listener who has not heard of this. Past this point when a viewer who has seen starts wondering about and trying to figure out what it all means I think it all starts with what each individual viewer thinks. I provide descriptions and I give my opinions on various topics related to a Return. Opinions expressed by individuals in the future concerning a Return will surly vary greatly with each other and with my opinions. I think that's just the way it will be and maybe that's how it's suppose to be according to human nature. Certainly the discussion will be without end as it already is stunningly enough. I'm hoping that seeing and realizing these things will cause people to pause and consider our future as a species. Peace and good health for future generations on a healthy planet is very desirable and appears to be our only hope to survive. We have to figure it out and get busy fixing the problems on our planet everyone working together, that's our only hope.

A very fantastic thing happened and is still happening!

At some point on the Earth life somehow starts. Then at a later point some of the life on the Earth looked up and sees the extremely well defined muscular human form within the Changed Light from the Moon. Then from that time forward through the history of life the Changed Light event has been happening and life has been looking up and seeing.

Clearly there's a basic question to be asked. Either what happens in the Changed Light does effect the life that sees it or it doesn't effect the life that sees it? Either there is a big full circle of human life happening here in direct conjunction with the actions of the Object and especially the human shaped form that is produced by the big main shadow's point interacting with the Moon's Changed Light and then life on Earth seeing that human shape or there isn't. I am unaware of any possible middle ground here on this question to be found or somehow realized. Once the Object has returned and caused this repeating celestial event to happen again these basic questions will be asked because it's going to be clearly obvious that these questions are there.

The basic question instead will be not whether or not human life is connected with the human shaped Changed Light but how did we humans happen? Why are we an exact copy of the muscular Man of Light? I describe exactly what happens and think that at a very deep fundamental level humans are here because of simply looking at that human shape during the time of our life form's earliest evolution. I am going to guess that there was the point in time when our first ancestor looked up and saw the muscular human form within the Moon's Changed Light and that was the point in time and the action at the root of our origin. Future researchers and scientists and people everywhere will be always finding new questions to ask and making discoveries as everyone tries to understand as much as possible about this entire event and human life. These facts are stunning and will be confounding for the world. It will be apparent to everyone that at a basic level this is what happened and because of this we are here and we are the ancient mystery.

The Changed Light has special characteristics. Suddenly one effect is somehow you start to see from up there as if out of a remote camera. This particular visual effect starts at the exact point when the crossing Object's surface suddenly looks flat and at the same time it starts to tilt towards the viewer. Now the viewer is finding that they are seeing the surface from very close. Look to the right at this point and the sight of the ancient Valley is there. The way the viewer sees from up close ties into and works together with the fact that the Changed Light's special qualities mysteriously causes it to imprint itself into the viewer. This imprinting action starts from the initial point in the Changed light event when the Valley is seen even before the forceful pressure that is physically felt when when the light mountain rises. It is my opinion that this effect is key in understanding how we as human beings down here on Earth are an exact copy of the way the Changed Light from the Moon looks and the form it is in after it was shaped and sculpted by the first one or three shadows.

Very suddenly the ancient Man of Light stands up and turns around and suddenly looks directly at the person who is looking face to face! I believe that the face of the Spinx is the face of the Man of Light.

Then later much later in time The Eagle, the Apollo 11 lunar lander carrying Neil Armstrong and Buzz Aldrin lands up there on the Moon the place that is the source of the light for the Changed and Lasting Light event. The place where the first initial sculpting and creating by the shadows happens ultimately the very place where we come from!

Neil Armstrong opens the door and climbs down the ladder of the Lunar Lander and then he takes that last step down onto the surface of the Moon! The ultimate ancient mystery is there! A man standing on the Moon! The entire weight and forward momentum of all life's history and even all of basically every kind of Earth and solar system history and the history of everything and everyone and especially the history of the awesome goodness of the primordial Man of Light himself picked up Neil and Buzz and set them down on the Moon. A human being then climbed down and stood on the Moon and spoke those words completing the ancient Big Full Circle!

And all along everyone thought it was basically just the team that worked through NASA that did this. Everything is way bigger than we realize!
There really is a truly magnificent awesome giant Big Full Circle that the ancient Man Of Light started and then Homo Sapiens flying Apollo made complete!

The Big Full Circle

"That's one small step for man, one giant leap for mankind."

Neil Armstrong

7 THE N.L.T.

New Living Translation Bible

I am going to describe a situation, a description in an ancient text that happens to be in the Bible and an ancient Lasting Light sight that I saw up over and beyond the upper right horizon of the celestial object I saw crossing down through in between the Earth and the Moon that are one and the same.

On this page I'm not going to be trying to prove anything instead I am simply going to describe and compare my own experience and some of the sights I saw during the Object's last return down across in front of the Moon with an ancient text that I know also describes the exact same incredible celestial event and in particular the exact same specific repeating special ancient sight seen within the incomparable Changed and Lasting Light from the Moon.

Again I will state that I know that it will take the Ancient Object's return and then the sights that are seen in the Moon's Changed Light to reoccur and then and only then before people are able to finally realize that I really am describing a situation that is real.

At a most basic level I describe the sights that I saw during the Object's last orbit down across in front of the Moon. Sometimes I am fortunate to find or notice ancient depictions and texts that were obviously done by people who also saw the same Object and the same repeating sights that the Object causes to happen in and then out of the Moon's Changed and

Lasting Light. For me this is actually one of the most interesting area of things for me because when I do find things that are done in ancient times by people who knew and maybe also saw like I did I feel an excitement that comes from the connection with these people that I know is really there. Everyone who looks up at the right time and then continues to look in an uninterrupted manor will also see the exact same sights and the exact same things that humans have always seen happen all the way back through time.

I am pointing out some of these things that I notice because I know that pointing out some of these instances will definitely help future Object and Lasting Light viewers. Before that time arrives many or most people who read from these pages will be certain that there's no possible way I could be describing in any way shape or form any sort of possible real experience from my childhood. Later people are going to know that the only way I could have possibly known about and accurately described exactly what happens during the Object's return home is to have actually seen the Object's last orbit down across in front of the Moon the last time the Object was actually here.

I have been asked this question; "Even if you did see the Object and the Moon's Lasting light as you describe it, that was over thirty five years ago so how could you possibly remember well enough to provide accurate reliable details of what you saw?" I am sure that and know that many people can remember the details of important events they were a part of or saw many years earlier especially if these events were spectacular, stunning and or traumatic and important etc. That seems to be true and along with that, for me along with always being known by my family for having a good memory concerning those early years I know that the reason I list next below is the main reason why I know the details of what I saw that night so well.

I think that it turns out that when the crossing Object's forces change the reflected sunlight from the Moon as seen from the Earth, somehow the suddenly differently appearing and behaving light from the Moon acquires an indescribable amazing and strange ability to imprint itself into the viewer. I now that I am able to remember what the Changed Light sights look like in an unnaturally clear way and I am even see the overall shape of the Changed Light again thanks to the mystery of the Shape Effects in a way that is not within normal experience.

Also this goes farther for example I believe that a person who has seen the human looking forms within Moon's Lasting Light is able to notice that they will be actually re experiencing the way they were suddenly forced to feel when they originally saw the Lasting Light later when they are remembering the Lasting Light. A truly bizarre real and unavoidable effect that I suspect future Lasting Light viewers will also know. I think that this effect just happens to you automatically when you see and is directly related to the Shape Effects and it's imprinting ability

Through the use of a small backyard telescope I saw the sights and sculpted human forms that the Object's shadows create in the distance of the Moon's Changed and Lasting Light in close up extreme detail. Concerning the part of the event that I saw I know it very well in very fantastic incredible detail. I am certainly not an expert in any overall sense but I know for certain that when people see the Object and the sights it's shadows sculpt and create using and with the Moon's Changed Light people will realize that my descriptions are actually real and accurate. If people do decide to criticize me they may say that I did not say more and describe more of the sights that are seen. It is true that so far I have not described everything that I can think of to try to write about however I do describe the overall scene and all of the main events that I am aware of. There is an incredible vast sea of details and describing all of these details is not possible. I point towards the entire event, the overall scene and specific details. As a central part of my responsibilities I provide accurate descriptions of the details of the major sights that are seen when the speeding forward rolling Object returns down across in front of the Moon and that's the point for me.

The ancient people were very focused on the crossing Object and especially the sights that Object's shadows create in the Moon's light after the Object's forces change the Moon's light and then what the Lasting Light shapes do after that as the viewer looks up through the layers of Lasting Light too see them. I know for certain that many ancient texts contain descriptions of the sights that are in fact made and sculpted and created by the Object's shadows in the Changed and Lasting Light from the Moon.

The points I am making and the descriptions I offer do in deed describe a real event that repeats and happens over and over again. Definitely the Object's return and the Changed and Lasting Light event that it causes is the big thing that happens. I know that it's important for people to know what happens ahead of time before the Object returns home so that at least they will be ready even if it's only in a small way. It's important that people have heard the of what happens between the Moon and the Earth so that they can also see some of the sights that the ancients wrote about exactly the same way that I also saw some of those fantastic ancient Lasting Light sights and anyone who looks up at the right time can see. In order to help you see the ancient Lasting Light sights you can trust the points and the details and the viewer's tips I offer in my Return Viewer's Guide. The Return Guide is as a very reliable dependable place for you to start from.

It turns out that the present world situation is going to be changing suddenly because of the fact that the ancient Object of old is always nearing and getting closer to the end of it's present orbit and soon it will be detected. The course of modern world and human history will be forever changed.

THE PRESENT GREAT YEAR IS ABOUT TOO END!

When it does end the powerful forward rolling Object will be back crossing down between the Earth and the Moon intercepting and changing the Moon's light once again.

Forget about today's comfort level and the way people are certain about what they believe to be true concerning the ancient texts they study. Among many sources I have checked out and also some religious study guides there is no mention of the crossing Object and there is no mention of the fact that the light that the ancient people adored was actually the Changed Light from the Moon and the Changed Light from the background of space. The latest most profound understandings of today and all the excepted truths that the scholars and the theologians and historians agree upon and point towards include no mention of the Object or especially the Changed and Lasting Light from the Moon in regards to what it actually really is as opposed to what people today think the ancient references to light are about and how it all happens. This fact does say a lot about the way today's understandings are completely totally lacking and don't at all begin to describe what actually really happens or even were it happens.

I've said this before but once again I can't begin to understand the size and gravity of the shock to come once the rest of you start to get up to speed and begin to realize the true situation. As if the reality of the Moon's Changed Light isn't going to be difficult enough I know for sure that once the ancient forward rolling Object is detected and is seen speeding down towards the Moon and our Earth again the shock will be multiplied many many times by terror because of the Object's very intimidating presence

I obviously do look forward to seeing the Moon's Changed Light again and especially the sight of the incredible fantastic very muscular ancient Awesome Good but I am looking forward to the the Object's return with a large amount of real fear and dread. I know enough to be afraid of the Object and I am. Very difficult times are nearly here. I do take great comfort in the fact that the Object's last orbit down across in front of the Moon seems to have happened in complete safety and everyone else can take comfort in this as well. The crossing Object's last orbit down was a safe return and this is a very important fact to be remembered once the returning Object is detected

I know that many places within the Bible contain descriptions of this real celestial event. I know for a fact that descriptions of the Object and especially descriptions of the Moon's Changed and Lasting Light are a main focus of the Bible. Real descriptions of a real and stunningly profound event.

The ancient people knew that the Lasting Light sights they looked forward to seeing were all important. It seems to me that seeing and especially seeing up through the many layers or levels of the oval shaped plain of Lasting Light to the ultimate special place deep up within the oval shaped plain was the main goal in life for the ancient people who knew and were familiar with the details of what actually really happens when the Object passes down across in front of the Moon. There seems to be an ultimate place to see from and an ultimate place to be made to feel from within the Changed Light. This is a guess of mine based on what I saw and what I can read in ancient texts concerning things that are seen and things that happen in the Changed Light after my view of the Changed Light ended. Seeing from this ultimate special place that I'm guessing about was a central main point of all of the ancient Return Viewer Guides that have been passed down through time to us in our modern day.

For most of my life since I saw the Object I never knew that I could read about the Object and the Changed Light in some ancient texts. I know that people in the future will think that this is strange because it will be so obvious to them in their day but this is something that I just simply didn't know even though I did see the Object and what happens to the light from the Moon. No one was giving me hints or pointers along the way and certainly unlike in the near future and in the distant past the details of a Return event just are not widely known today never mind understood. Also the fact is that I never took the time to read any ancient texts of any sort because I hadn't thought to do so and one reason for this is the connection between religion and the Moon's Changed Light was not apparent to me.

Also like I have touched on before I just didn't think about the Object very much sometimes because unfortunately I was unable to remain thinking about it for very long before suddenly I just simply wasn't thinking about it anymore. Like I have stated elsewhere I do feel bad about that situation but that's just how it was for me for many years. I also talk about how I slowly overcame that situation that I was in and eventually and since that time I have learned and taught myself how to not only think about what I saw but I have learned how to focus and concentrate on the memories of these sights that I saw.

In the big picture it does turn out that being able to concentrate on these memories is central and basic and even all important. The ancient people certainly knew this and they developed this ability into a highly refined art. Unfortunately it took me thirty plus years to fight back up through the complete shock of seeing the Object and especially the Lasting Light before I could begin to understand and know that concentrating on how these memories look and how these memories make you feel is part of the basic big important point of it all.

I feel very fortunate that I now am able to realize that I have this ability. This will happen and does happen in an unavoidable way to a person if they see the Lasting Light. If no one points this out to you, you will either figure this out for yourself or you won't. There is no guarantee and I know a person could see the Lasting Light without later realizing that they can actually see the sight of the Shape Effects. I know the Shape Effects at least that's how I think of and try to describe the Changed Light's imprinting effect. I am very lucky. I can see the horizon come to life and even step beyond if I really try. If you see past the Object's upper right horizon when the Object is here you really can see from up there again later on even many years after. A fantastic effect and I have actually found myself reading about and I am able to recognize this effect thanks to some ancient writers. I know for a fact that they saw the Changed Light and exactly some of the same sights that I saw.

I saw through a telescope, I don't know what the implications of this are but I know that I saw the details within the Changed and Lasting Light from extremely close in a very fantastic way and this can happen for you as well.

Over the years I looked up and read many stories and versions of stories and myths and legends etc. I found a reference to mountains of Atlantis and also a special most important mountain. A mountain that had a summit inhabited by the Gods. Then I also found it suggested that many ancient religions and even some modern religions had stories about a special or sacred or holy mountain of some type. This was actually a shock for me because I simply never knew this before.

I have described the mountains around the perimeter of the ancient valley elsewhere but as a way to try too help future viewers I though I would take a moment to mention here about two very special mountains located straight in the distance of the valley that I don't think I have described or written about before. I have always though of these mountains as the two mountains of the valley. These two mountains are special at least to me because I looked and saw between them further into the distance deeper into the Lasting Light. With the telescope I was using I found myself able to see very far and I was drawn there the way the Lasting Light draws you forward too see. I think that the two mountains are another fantastic sight to see and notice. I believe that at that point I was lucky to stumble into what I think may be a clue to where to look in order to find the correct path to follow later up through the layers or levels of the plain. This is another area where I try to guess based on what I saw. I have a feeling that the two Mountains may possibly be key and possibly all important later for future viewer's who hopefully see further and deeper than I ever did. Remember the Two Mountains!!! Look between them and be drawn deeper than I saw from. Check it out and decide if this is the way. I don't know for sure but always remember that you can see between the two furthest mountains located straight ahead approximately centrally in the valley. This may or may not be important. If it is important then chances are this is all important and nothing less. Even if the two mountains of the Ancient Valley aren't the way to see they are still special at least too me because I think that seeing between them might have been the furthest place I saw.

As a part of my own personal family tradition as a gift to my two children I named the ancient valley of old after my daughter and my son. S & W's Valley. I have exercised my ancient human naming right to do so. A right that does not effect or infringe upon anyone else's rights. Naming the ancient valley after my children does not change the ancient name or names of the valley as far as any of your ancient traditions or customs are concerned. On my Naming Rules? What Naming Rules? page on my website I talk about everyone's equal rights. I also talk about how everyone's ancient traditions customs and beliefs are equally valid. I know that the sight of the Awesome Good and all of the Lasting Light sights including the mountains of the ancient valley belong to everyone equally no matter were they live or when they lived. Anyone can name the Object or any part of the Lasting Light event. I named the ancient Valley after my two children. I have chosen too use the Object's ancient name Adoil. I have the right. It's an ancient right that's mine. You have the same ancient right, a right that you are entitled to because you just like me and everyone else have ancient ancestors who saw the Changed Light.

Like I mentioned I read that many ancient religions and even some modern religions had stories about a special or sacred or holy mountain of some type. During the Object's last return I saw many mountains. Without any question one of these mountains was definitely very special.

I believe that in ancient times the entire Lasting Light event itself progressed further than it did during the Lasting Light event that I saw. When the plain of Lasting Light emerged out of the outer layer or the tunnel or sleeve of rising Lasting Light many more types of mountains were possibly seen. Also the sunken inner top of the rising mountain of Lasting Light complete with the Awesome Good and the other figures down within was surely also a part of what the ancient viewer was able to see within the later plain of Lasting Light at least at some point. It sure sounds that way to me and I feel this way because of some very stunning things I have read that I know for sure are ancient descriptions of the Moon's Changed and Lasting Light.

Going back, after reading about the important central mountain in Atlantis I was suddenly struck by the thought that I might be able to read about the Object and more about the Moon's Changed and Lasting Light thanks to the fact that the ancient people had written many things down.

Again I know that this is just to obvious and I somehow should have figured this out a long time ago but that isn't what happened for me. If a person does not know something then that's how it is for them. It's sort of like how you all didn't see and you don't know what happens. I keep saying that it's not your fault and it isn't your fault. I think it's about time I started to feel the same about myself. I did see the Object's last return and I did see the Changed and Lasting Light from the Moon. I was not able to understand it and start to figure it out for a very long time. I know that that situation wasn't and isn't my fault.

As result of that situation it took me a great deal of time to realize that I could read about what I had seen happen between the Moon and the Earth in ancient texts today.

I am going to describe a situation, a description in an ancient text and a Lasting Light sight that I saw that are one and the same. The ancient sight that I am going to describe repeats over and over again in exact detail every time the Object orbits down across in front of the Moon and the Object's shadows sculpt and shape the Moon's magical Changed and suddenly Lasting Light.

One day everyone will know for themselves that the situation and example as I describe it does exist and is real and I am right and correct in the association and the connection I have made and am going to describe below.

In the end people will know this for themselves because they will have read the same ancient text and they will have also seen this fantastic ancient Lasting Light sight and they will know with the same certainty that I have that there's nothing else that the situation and example that I'm going to describe could possibly be. I am going to describe a situation, a description in an ancient text that happens to be in the Bible and an ancient Lasting Light sight that I saw up over and beyond the upper right horizon of the Object I saw in between the Earth and the Moon that are one and the same.

I have had a Short Wave radio sitting on the shelf for quite some time. I used to use it to try to listen to distant news broadcasts. I was not always successful locating and tuning into the news from distant countries but I was always very easily able to find many radio programs that featured various types of religious broadcasts. I think this was the time when I first started to realize that there were a lot of terms and themes being talked about on these broadcasts that were very familiar to me and reminded me about certain things that I saw when I saw the Object and the Changed Light. This was an exciting time of new discovery for me as I found myself suddenly hearing about things seen in the Changed Light!

At one point I had the SW radio out and over the course of a number of weeks whenever I had time I scanned the dial trying to tune in these various religious broadcasts. I found listening to SW radio to be very hit and miss and unsatisfactory. Once you find a SW radio broadcast you have to contend with the fact that many of these signals fade away. More expensive SW radios have a signal lock on feature that my radio lacks. Also I very quickly found that along with hearing little actual text from the Bible I found my listening time filled with much commentary and a great deal of guessing as to what it all means. For me the text is the interesting part and not the commentary that tries to explain what it is all about and what it is suppose to mean. It was not long before I realized that it was time for me to go out and get myself my own copy of the Bible and that's what I did.

My first Bible was a copy of the New King James Version, ABS. For the first time I had my own copy of the Bible. This is when I first realized that there were different versions of the Bible. I searched second hand stores and used book stores and gradually I started to acquire other versions of the Bible to add to my slowly growing collection.

On this page I am going to describe one of the Return descriptions found in the Bible. I plan to put this ancient description in context describing how it fits into the entire first part of the Lasting Light event. Were it happens and when it happens when the Object rolls down across in front of the Moon. By doing this I will be showing how this Lasting Light description in the Bible is incomplete. However although this description is incomplete at the same time this Bible description does accurately describe an important segment of a central most important event exactly the way it really happens!

The specific Lasting Light shape that I am going to describe repeats over and over again in exact detail every time the Object orbits down across in front of the Moon and the Object's shadows sculpt and shape the Moon's Changed Light. The bottom lime is that incredibly it's all about what happens and what is created when the shadows that are cast by the Object's Standing Stone style craters interact within the Moon's Changed Light, a very very fantastic thing!

One day everyone will know for themselves that the situation or example as I am going to describe it does exist and is real and I am right and correct in the association and the connection I have made and am going to describe below. In the end people will know this for themselves because they will have also seen and they will know with the same certainty that I have that there's nothing else that the situation and example that I'm going to describe to you could possibly be.

As it happens there's only one Object that comes to pass down in between the Moon and the Earth. Shadows are cast from all of the Object's many vertically standing crater rim pieces. Each of these individual shadows travel into the Moon's fantastically Changed and suddenly Lasting Light.

This event repeats in exact detail every time the Object crosses down in front of the Moon. The same sights are seen by anyone who happens to look up at the Moon at the appointed time. I saw the speeding Object cross down and the Lasting Light sights through a small backyard telescope. I saw these incredible confounding sights from very close up in extreme detail. One day people everywhere including you are going to know and understand why I freely describe distribute and share the details of what happens so that you will also know so that you will have the best chance possible to see the Object and the Lasting Light sights with foreknowledge.

The ancient people saw these sights and wrote about what they saw. The Bible and many other ancient texts contain many of these descriptions. I have noticed that often these descriptions are usually incomplete and out of order in most cases and also intermixed with all sorts of commentary added at various times. Sometimes this commentary is very relevant especially if the source for this commentary is the original ancient author. As far as I can determine often times various commentary within the texts is seemingly unrelated to the actual facts of the literal sights being described by the original ancient writer.

I know that because my overall understanding is limited I just don't know how much of the commentary found in the Bible for example is actually related to what an observer experiences and feels when they see the Lasting Light event in it's entirety. This is just to big of a question for me to be able to understand and answer with certainty. Even with the unrelated commentary many ancient texts including the Bible contain descriptions of the Object and especially descriptions of the Moon's Changed and Lasting Light and the entire event.

This brings me to the specific special Lasting Light sights that I saw and want to describe here and the ancient written description of this Changed and Lasting Light sight that is found in the Bible. I have described the incredible sight of the muscular back and the right side of the ancient Man of Light, the Awesome Good in many places and I know that this incredible human shaped form is described in the Bible.

After a while I discovered The King James Version audio Bible online. Over the course of many months around the house I would listen to one book after another virtually all of it new to me playing in the background as I went about my business. Obviously the Bible is a very fantastic and very special ancient text. Since that time the more I learn about the original roots and the original way the texts reads the way the ancient people wrote it the more fantastic it is to me.

While listening to the Authorized King James Version Bible online I heard this;

Exodus 33 (King James Bible)
21 there is a place by me, and thou shalt stand upon a rock
22 And it shall come to pass, while my glory passeth by, that I will put thee in a cleft of the rock,
23 And I will take away mine hand, and thou shalt see my back parts: but my face shall not be seen.

Really? Wow I remember the moment that I heard this for the first time. It was as if the words "thou shalt see my back parts," jumped right out at me!

Somehow the crossing Object changes the light from the Moon while it is between the Earth and the Moon. This is what really happens but I have no idea how this effect actually happens. The Object's vertically standing crater rim pieces cast their own individual shadows to the right towards and then into the suddenly Changed Light from the Moon. Somehow the Moon's Changed Light is fantastically shaped and sculpted by the shadows.

The very very black darkness of the shadow really does travel or lead to the Moon's Changed Light just like it says in the Bible. Darkness leads too light. and we can all be thankful that it does in fact do this because of the fact that somehow some of the life on Earth that looked up and saw these Lasting Light shapes and forms that are created by the shadows was somehow caused to copy and model itself after these forms or shapes that are light. There is nothing simple about it but somehow in the end it really is as simple as the last statement that I just made.

The ancient Valley is the first Lasting Light sight that is seen to the right and upper right compared to the Object's position when it's obscuring the Moon's disc. As I describe in my Return Viewer Guide the shadow first appears under the overhang that develops on the forward rolling Object's raised surface area. The shadow spreads and grows under the overhang until it finally very suddenly drops down to the Object's surface below.

The shadow then instantly starts moving very rapidly across the Object's surface towards the Object's upper right horizon through a sea of circular craters. All of these craters are ringed by vertically standing rim pieces. As the shadow rapidly moves a group of tall narrow individual shadows emerges ahead beyond the edge of the main wave of the shadow. These are the shadows that do the first important sculpting and shaping and creating in and with the Moon's Changed Light.

A central most important group of three main shadows becomes most prominent and obvious. The middle or central shadow is taller and twice as wide as the two smaller outer shadows that travel with it one on either side. I know that this central biggest shadow is the ancient Black Road and it's all important to focus on this shadow's point as it rapidly moves away from your viewing position. If you do this you will find that you are starting to suddenly see from up there as if out of a remote camera an effect that actually starts happening to the viewer from the first instant that the Moon's Changed Light becomes visible. A very special incredible effect that just simply happens to you if you are looking. Suddenly the point of the big main shadow looks as if it's a flat ribbon as it starts to rapidly go up and down and up and down while it is moving away from you deeper into the distance of the flat white clouds of the valley floor. Then instantly suddenly the point of the shadow starts to cut back and forth very rapidly drawing a line that goes all squiggly zig zagging seemingly in every direction.

The fantastic very magical sculpting and the shaping has begun as you watch this split second in time. This is the sculpting and shaping that causes the very muscular looking back of the Awesome Good to be created along with the shape of the entire Awesome Good sitting or crouching at this point facing away from the viewer towards the extreme distance of the valley!
At this point in time instantly suddenly everything that you are seeing has changed. The place were you are seeing from is no longer racing

forward anymore. The entire scene that you are seeing has changed as well. Now you are actually looking out from a place that is deep in the distance and slightly below the level of the flat white cloud effect that is the valley floor.

Your viewing position is within the top layer of the flat layer of mist. Also you are stationary not moving forwards or backwards. I don't know if you can look in every direction at this point or not. Either I looked down in front of me by luck or the place down in front of you is the only place you are able to see at this point. I am able to realize that it is possible that I might have seen down in front of were I was seeing from by luck. Also I know that at times you are only able to see were the Lasting Light allows you to see so I am not sure what all the possibilities might be at this point. Definitely you really need and want to look down in front of were you are seeing from at this very special point in time within the Lasting Light.

Right now I believe that you are now seeing from the place within the Changed Light that is described by the ancient text as being, "a cleft of the rock." At this point you are now looking down from a place that is located close above and close behind the sight of the very muscular back and right side of the fantastic ancient Man of Light.

I saw this because I happened to look up at the right time and at a very basic level it really is as simple as that. Anyone who looks can see this sight if they look up at the right time and they look at the point of the big main shadow and then they keep looking in an uninterrupted manner. I was using a telescope and we were looking at the Moon originally. Suddenly the Object was just there rolling, the shadow dropped and the shadow's big main point drew and sculpted and shaped and created the big sight of the Awesome Good as I describe. Truly an incredible fantastic thing! Incredible!

It's no wonder why it took years for me to be able to realize that I could actually read about what I saw in a book such as the Bible. Any time I had ever heard or read anything about the Bible or any religious text or religion for that matter no one ever mentioned anything or pointed towards anything specific that had to do with or was even close the literal sights and the entire celestial event that I had seen up there between the Earth and the Moon!
Unfortunately today's understanding concerning the details of what actually really happens is so far off from what actually really happens

that the fact is there is no understanding in the world today concerning the details of what actually really happens. And people are wondering what's the matter with me when I say I'm worried about the size of the shock that's about to come.

I always thought about seeing that first glimpse of the Awesome Good as seeing from above and behind mainly because that's exactly how and were you suddenly find yourself seeing him from, above and behind.

KJV
Exodus 33
23, and thou shalt see my back parts:

I was very excited and even blown away when I first heard and then read this passage! Right away I knew exactly what was being described. Although incomplete and although the overall description is not entirely clear to say the least I still knew exactly what the moment was and exactly what the sight was that the ancient writer was describing. But "back parts," is not how I would have described this sight but through the translation the meaning is still clear. At first after the big main shadow strikes and creates in the Changed Light the muscular back of the Awesome Good is what you see.

Translation issues cloud and confuse many more times than I know plus translators seem to inject their own versions of meanings into the texts. It's to bad that this has happened but I suppose it's unavoidable for translators who have no idea concerning the literal sights the ancient writers were describing but instead have all sorts of various different meanings and beliefs in their heads. As I mentioned earlier I used to search the shelves used book stores for any books that had ancient subject matter as well as for Bibles. If I saw a Bible that was a different version other than the versions I had so far I would consider buying it depending on condition and price. Importantly and routinely when I checked out a different Bible version I would and still do check certain books, chapters and verses to see the results of the various translation attempts.

One morning I was checking out some used books and Bibles and I found a copy of the N.L.T. The New Living Translation Bible.

As usual I started checking out certain places in the Bible that have meaning for me because I know with complete certainty that these places in the text do definitely describe sights that the viewers sees when the Object returns and changes the light from the Moon. At some point I checked Exodus 33 and I went straight to verse 23 as usual. Much to my surprise I read this; "you see me from behind"

Fantastic! This is how I have always thought of this sight!

One thing I have to wonder about is the idea of the way this is written. Who is suppose to be actually saying, "you see me from behind?" I saw this sight and I certainly didn't hear any sort of great booming voice saying anything. Who do people think is suppose to be talking or writing here? I can't say or guess what people think about this part of the text and all of this text for that matter. Overall that or this does describe the situation for me. Basically I don't know or understand how or what people think about this or that passage of ancient text. Generally what people think about their ancient texts is of interest to me. Like I have stated elsewhere, I don't know what you believe and I don't know how you feel about what you believe. For me what people think and believe is important but outside of my area of things and too big of an area for me to consider. I know the sights that I saw and I find descriptions of these sights in many ancient texts. I am pointing this out for people who are interested and also I am pointing these examples out as a way to be honest and true concerning the details of the extremely important sights and things that happen in the Moon's Lasting Light.

The truth of the Lasting Light is important and within my area of things and these are the details that I know first hand. Again the size of the shock to the people of the world that will happen eventually is way bigger than I can imagine and then there's me trying my best to do the right thing in this time preceding the crossing Object's next orbit down People will be left to deal with these facts and figure this out for themselves. Certainly people will struggle with this but the truth of what happens in the Changed Light is so profound that it will be clear to people that there is a big thing that happens and this is certainly it so with those facts as a starting point I think that people will able to take their core values and beliefs and see that there is no contradictions and the goodness of their religion for example with be strengthened.

Overall at a basic level my job is simple. It's my job and my responsibility to describe what I saw and that's what I do. I have also decided to point out ancient some ancient Return descriptions that also describe the same sights that I saw. Again repeating myself I have said before and like I keep saying, I can't imagine the shock to come when people are finally forced to realize for themselves that the Object and the Lasting Light event as I describe are in fact a reality because the Object of the Crossing Down has returned and has forced them to realize the true situation. Once the big main shadow strikes and many people see the muscular back of the Awesome good for themselves there is going to be no denying the facts and the reality of what really happens between the Earth and the Moon. Also people are going to know just exactly the same way that I do that the Bible does indeed contain ancient descriptions of what happens when the Object crosses down in front of the Moon.

Exodus 33 (New Living Translation)
22 As my glorious presence passes by, I will hide you in the crevice of the rock and cover you with my hand until I have passed by.
23 Then I will remove my hand and let you see me from behind. But my face will not be seen."

After the the ancient Man of Light suddenly gets up and turns around to his left to face you directly. A central point from that second forward is the idea of remaining locked into the incredible face to face ultimate staring contest for as long as possible. During the time that the second phase of the Changed Light event when the Light Mountain is rising you can look away from his eyes and you can look at the other human figures that are arrayed on both his sides and around behind him or you can look around at the entire scene as well while still being able to look back at his eyes. You can look away from his face and look around during at least the first six or eight seconds or so of the rising Light Mountain. After that point my view ended. From what I am able to gather from descriptions that I have found basically the viewer is suppose to remain looking into his eyes going face to face with him and I think that is true. While that is true and I always say don't look away you don't have to look into his eyes 100% of the time. As it turns out this looking into his eyes is extremely intense at this point especially along with the way the pressure that flows from the Changed Light is being felt at this point. The most critical consideration for the viewer during the entire time they are able to look at the Changed Light is to never look away from the area of the Changed Light and instead for example turn to the next person standing beside them to make comments and conversation! Very bad idea! Again a basic fundamental point is that during the entire time of the Changed Light event up until when the Light Mountain rises you are able to see this from at first close to the Object and then ultimately you can see from within the area of the Changed and Lasting Light. For me I was looking in the telescope and in the view finder and I was seeing an image that was exactly the same as looking out of a camera located very close to what I was looking at. Very strangely the viewer is now able to look in a different direction from the point were we'll say that the remote camera is located. This is the effect that happens. I didn't stop looking at the area of the Changed Light and then remove my eye from the telescope eye piece and then look back again to see more so I can only speculate and guess about exactly what would happen. My feeling is that generally speaking you would have to start over again looking at something and then seeing from there. You simply are not going to see as much and you might not be able to see from as deep within. Hopefully the person standing next to you is also looking and then later hopefully they will still be there after it's all over, you can look at them at that point. Don't lose your opportunity to see and instead waste your time looking at the person you are with while the Changed Light event is on! The idea of how you feel and the idea of remaining looking because that's what you are suppose to do is important and hopefully helpful for future viewers who will be badly struggling at this point. The face of the Awesome Good is seen! His penetrating stare is over powering and completely extremely difficult to describe. The incredible big face to face! The Big Noticing does really happen just like certain ancient texts accurately describe! Forget about how this moment is thought of to be symbolic of this or that abstract imagined idea or meaning related or unrelated. The Awesome Good gets up to his left and goes face to face with the viewer and once again it's as simple as that. This is what really happens!

Exodus 33 (New Living Translation)
23 Then I will remove my hand and let you see me from behind. But my face will not be seen."

This does describe exactly what is seen at the initial point when the Man of Light is first seen after being created by the point of the big main shadow. When you first see him you see his back. He is oriented facing in the same direction that you are facing towards the distance of the Valley, basically the Moon. You can't see his face until he has finished standing up and turning around and that's when you first see his face.

22... I will hide you in the crevice of the rock and cover you with my hand until I have passed by.

23 Then I will remove my hand and let you see me from behind

Somewhere lost in the translation is the part about how you see the muscular back of the incredible Awesome Good from above and behind and not just from behind.

You see down from above and behind the Awesome Good. Eventually someone is going to find a reference to the higher correct viewing position within an example of some ancient text. I am able to realize that within some original ancient texts this information is very likely there to be interpreted and then translated correctly. The ancient people describe these literal sights as a part of the texts they left behind for us. The information is there and it's only a matter of time before someone finds a reference to the part about how you see the back of the Awesome not just from behind, but from above and behind.

I can speculate; *I will hide you in the crevice of the rock" .... I will... let you see me from behind.*

It does appear as though the place that is the crevice of the rock is the place where you see his back from and you do see down from there so perhaps the place above where you see down from does appear in the translation just not in a plain clear way compared to where the text states "you will see me from behind."

As the glory of the Object passes by down across in front of the Moon the overhang develops shadow drops from under the overhang and then travels up and past the Object's upper right horizon into the Moon's Changed Light. A very special tall wide obelisk shaped pointy individual shadow that was cast from one of the many vertically standing crater rim pieces on the Object's surface suddenly sculpts and shapes the Changed and Lasting Light from the Moon into the incredible sight of the muscular back and arms and shoulders and head and neck and right side of the very very incredible magical fantastic Awesome Good, the ancient Man of Light!

Seeing and going face to face with the Man of Light is not a simple thing in that how you feel and what you are thinking when you see is very important in my opinion. As a part of the goals I have set for myself I want to help future viewer's as best as I am able. Seeing the Awesome Good suddenly stand up and turn around to face you can cause fear to be felt. This not how I believe it's suppose to be but fear experiencing fear at this point is surly an inevitable part of seeing him.

I make an effort to help future viewer's to not feel fear but the opposite instead. I know that this effort is only going to be partially successful. Fear is an inevitable part of seeing the Lasting Light sights. I believe that this is probably just the way it has always been all the way back through ancient times. Along with knowing this first hand for myself I have read many ancient accounts of how seeing and fear go hand and hand. It's important that people know that it's more than just fear and that the ultimate goal of the ancient people was to see face to face and possibly eventually also see up through the layers of the Lasting Light of the oval shaped plain to the ultimate place in the Lasting Light to see from.

I have read that this is the place of ultimate peace and love and joy and grace. The hook-rays are there and it appears that this place has a special colour. I believe that the ancient Egyptians might showed the hook-rays as the light rays with the small hands. Fortunately the viewer does not have to see from as far and as deep as this before they can start to know the love part of it all. I think that from the time that you find yourself seeing down from above and behind at his muscular back concentrating on the idea of love and especially the idea loving the Awesome Good back from that point through to when the Awesome Good gets up and turns around to his left too go face to face with you and beyond you should begin to know the love that's there to know and not just the fear that's instantly automatic. Then the idea of remaining looking and not looking away is all important!

In this chapter I described a situation and a description in an ancient text of a fantastic incredible Changed and Lasting Light sight from the Moon that are one and the same.

8 NOBODY LIVES THERE

In spite of what many people believe today about "The Planet Of The Crossing," it turns out that nobody actually lives on the speeding ancient celestial Object that rolls down and orbits across in front of the Moon.

I am going to try to describe some of the main reasons for the confusion and complete misunderstanding. Contrary to what many people believe based on interpretations of certain ancient texts nobody lives there. That said incredibly the human form is seen there not just once but many times.

Using a telescope I saw the Object Of The Crossing Down and it's an incredible barren desolate crater covered world guaranteed. Just like when you finally see the speeding forward rolling Ancient Object first hand for yourself, when you are looking at the Moon from an up close perspective you are able to clearly see what's there. You are not going to look at up close images of the Moon and wonder to yourself does anyone live there?

Although the standing stone style craters on the Object are very different from the craters on the Moon, the moon coloured Object looks basically, exactly like a giant forward rolling fast moving moon type of celestial object. When you see the Ancient Object and eventually it will be from up close you will easily be able to clearly see the Object's surface details exactly the same way that I clearly saw the Object's surface details.

I believe that it is very possible that on May 24th 1926 German amateur astronomer Mr. W. Spill saw and reported the Object that I saw. I think that I saw the Object on the night of May 25, 1972. Mr. Spill called the Object the Earth's second moon. He called it our second moon for a number of reasons. One of those reasons was because that's exactly what this place looks like a moon, a barren desolate world with craters. The Object's colour matches exactly with the colour of the brightest areas of the Moon. The Object has the Standing Stone Craters instead of the lunar craters that we are all familiar with and the Object's speed and motion and orbit is completely different than our Moon's but the Object does look like a moon.

Did Mr. W. Spill see the Object change the light from the Moon? I have no idea and I would like to know the answer to this question. In the meantime Mr. W. Spill did see the Object "obscuring the Moon's disc," as he describes and certainly he didn't see anyone actually living there either. Actually I don't know the exact circumstances but I wonder how he could have seen that the Object obscures the Moon's disc without also seeing the way the Object's forces change the light from the Moon. I do know that since the Moon's appearance becomes drastically changed it is entirely possible that Mr. Spill could have seen the ancient Valley to the right and the upper right of the Object without realizing that he was looking at the changed and Lasting Light from the Moon. In Mr. Spills report he describes a black colour. Because of this it appears as if his reports' credibility or authenticity may have been called into question. An object in front of a sun lit moon should also be illuminated. It was three night before the full Moon. I think this was a basic problem for Mr. Spill's report. "a black ball," or "a dark ball," makes perfect sense to me because the big shadow that is cast when the Object's mound area rolls forward and becomes an overhang causing the shadow to form turns the Object's surface Black. Along with Mr. Spill's reported flight path matching the flight path as I describe his description of a dark or black object is not definite or certain confirmation but is a good indication that we are both reporting the same astronomical object.

I know only to well how Mr. W. Spill might have seen the shapes and human looking forms that the Object's shadows sculpt, shape and create with and out of the Moon's Changed Light but thought it best to not mention anything about it to anyone less they think that he was not credible. Sticking to the idea of describing the hard rolling ground of the Object itself can seem to be the best approach to take and that's what Mr. W. Spill might have done. If he did I don't blame him at all. That's exactly what I unsuccessfully tried to do as I describe in the first chapter.

Mr. W. Spill may have seen the Object and he did describe the Object as the Earth's second moon and not a type of planet that supports life. It looked like a moon to him just exactly the same way it looked like a moon type of place to me. As we all know a moon is not the sort of place that typically supports life, never mind anyone actually living there. When people see what is sometimes refereed by some people as the Planet of the Crossing nobody is going to be wondering if anyone lives there.

Today a very confusing situation does exist for some people. I can clearly understand some of the reasons for this confusion. Today some people think that some sort of humanoid life form of some type superior to us humans lives on what has been interpreted from an ancient text or texts as being the "The Planet Of The Crossing." Also these people sometimes refer to the returning orbiting Ancient Object as Nibiru or Nubiru, the Missing Planet or Planet X and many other names etc. I like and sometimes use the use the ancient name Adoil for the Object. I have yet to notice anyone else using the name Adoil for the ancient Object of the Crossing down however I am not surprised by this fact. These days Theia is my favorite name for the Object. I think that Theia kept going after striking the early Earth causing the Moon to form and today it continues to orbit across down through the solar system between the Moon and the Earth along it's original orbital path and is the Object of the crossing Down, the same object that I saw.

Without realizing it some people mistakenly sometimes think of some of the Lasting Light types of human looking forms that are written about that do indeed come from over and past and beyond the Object's upper right horizon towards the Earth as real living breathing humanoid looking superior beings, the Anunnaki for example. These human looking forms are real but they are not breathing and alive the way we normally think of a living breathing entity. These forms consist of or are made of Moon light that has been changed and made to last by the power of the Object's forces and then sculpted and shaped by shadows into the forms that are eventually seen by the viewer. As unlikely as that sounds eventually everyone is going to be suddenly forced to realize that this is actually somehow true.

Also by the way some ancient texts sound to me when I read them I am going to list some other ancient names that appear to have been used to describe some of the many very human looking lasting light forms that the shadows create. This list is only a small sampling and not complete and may or may not be entirely accurate; Anunnaki, Neter, Nefilim, Nephilim, Anakeim, Anakites, Anakim, Elohim, Nordic, the giants basically and so on. Some of these names may be the same with a different spelling I just don't know.

It seems that the name Anunnaki almost certainly was used in ancient times to describe the forms within the Moon's Lasting Light this much is seems to be true at least as long as the name Anunnaki is translated correctly. Once again some people think that some sort of a humanoid life form of some type, one that is superior to us humans lives on The Planet of the Crossing Down.

This is just one example of many many types modern day interpretations that happen to be totally completely and in this case even hopelessly off as far as the basic meaning and understanding of what really happens when the orbiting forward rolling Ancient Object Of The Crossing down between the Earth and the Moon returns is concerned. Many people get the sense from various ancient writings that there really is an unknown celestial object of some type out there that is an orbiting member of our solar system that has yet to be recognized, described and studied. Give these people credit because they are correct. Trying to imagine how an object you haven't seen may orbit through our solar system is going to be tough. The Ancient Object was crossing down between the Earth and the Moon May 24th, 1926. Now I think I saw the ancient Object May 25, 1972 and it could be back again May 26, 2018.

If a person hasn't actually seen the Object's return and what it does to the light from the Moon for themselves then they could never guess correctly what actually really happens as evidenced by all of today's incorrect guesses and complete misunderstandings. Today's ancient history experts and others know many fantastic ancient facts but as it turns out just like everyone else today they have no idea concerning the details and the facts of the Object's return to it's crossing point down between the Earth and the Moon and the way it intercepts and then changes the light from the Moon. Today's ancient history experts try their best but they simply are unable to describe a Return by guessing and trying to interpret meanings from ancient texts concerning what actually really happens. To start with the Object the Ancient people were focused and fixated on crosses down in front of the Moon. It intercepts and changes and then shapes the Moon's suddenly drastically changed and bizarre behaving light. These facts are completely unapparent in the high tech dark ages of today.

Researchers need a reliable place to start their research from. The Ancient Object crosses down between the Earth and the Moon. In spite of the fact that I just described what really does happen most researchers can only reject my description. They have to reject my descriptions because at first they have no other option. The details of what the Object is really like and what it really does to light is just to improbable sounding even to me. Like I have said before I understand completely and if I hadn't seen the Object and how it changes the light from the Moon myself first hand then I would have serious doubts as well in fact there's no way I would be able to believe it.

It does turn out that today's best guesses and ideas are completely off and come nowhere close to describing Adoil or Theia or the Ancient Object or whatever you want to call it and what happens to the Moon's light. The Object and the Lasting Light event that the Object causes is the subject of many of the most spectacular ancient texts that I have found. It turns out that I am easily able to understand that the ancient people saw the Object and the Moon's suddenly spectacular Changed and Lasting Light. I saw the Object and the Moon's Changed and Lasting Light and that's how I know for certain that they did as well.

I know that everyone's ancestors saw the sight of the spectacular form of the incredible Awesome Good. The muscular Ancient Man of Light. Once a person knows this from seeing these sights then it's easy for them to appreciate the fact that humans are all equal and all modern. We should be giving all of the ancestors of all of today's people the credit they deserve. Our ancient ancestors knew that seeing the Awesome Good and the other shapes that were also Lasting Light was good for you and that's how we all eventually somehow came into being. The Lasting Light muscular human shape from the Moon caused and causes the life that sees it to become exactly like it. The sight of the fantastic very muscular Awesome Good makes this an obvious conclusion, an obvious fact for me at least. This is a guess, my opinion.

Being separated by geography and time made and makes no difference because a Lasting Light event happens and repeats in exact detail every time the shadows that are cast by the Object sculpt and shape the Moon's Changed and Lasting Light. Unfortunately today with the world's lack of knowledge when a a person is guessing about what the ancient people were writing about and talking about they simply have no chance to guess correctly if they are talking about the Object related area of research or study.

I saw Adoil the incredible Ancient Object that crosses down between the Earth and the Moon. I'm not guessing about what is seen and my Basic Flight Path image is very accurate. Eventually slowly or perhaps instead very suddenly and dramatically, everyone is going to be forced to realize the true situation as I know it to be. The Ancient Object is near and so it won't be very long now before the shock of it's existence will strike and then at that point and beyond there will be no denying the facts of life such as they really are in our solar system and eventually especially there will be no denying the Lasting Light facts of what really actually happens between the Earth and the Moon when the speeding Ancient Object returns and crosses down.

Here's a couple of points about the confusion and misinterpretation that I see concerning sights that are seen within the Lasting Light.

A "second horn," among ten horns for example is seen as a part of one of the shapes that a shadow created by sculpting and shaping within the Lasting Light. K.J.V. Revelation 13, 1

Somehow according to some people, "the second horn is Iran." What? This type of thinking is so far off from the meaning of the text that I sense that it's even dangerous. There is nothing symbolic about the second horn and many many other similar types of examples in the bible. The people who wrote these ancient Return descriptions were simply writing down and reporting what they had seen within and as a part of the Lasting Light event as best as they were able to describe! Soon everyone is also going to know that this really is the case. For example; Many depictions left to us by the ancient Maya people are closely related to the Rev. 13, 1 Lasting Light descriptions because the Maya depictions I am referring to were crafted by people who witnessed the exact same sights. Read the text carefully and allow for inaccurate translations and different perspectives and frames of reference and then use your imagination. Once people are able to realize that all the ancient people were seeing and describing the exact same ancient fantastic celestial Object and indescribable incredible Lasting Light sights then true understanding may finally begin to take hold.

To suppose that "the second horn," or this or that other Lasting Light sight is somehow symbolic of a country or this or that in our modern day is not correct. Also to suppose and believe and teach that any modern country is pointed towards by any ancient Lasting Light description is to show a level of total misunderstanding and lack of knowledge that is truly shocking and troubling and unfortunate for many reasons. One day in the distant future well past the Ancient Object's next safe Return when the shock of the truth has had a chance to subside people in the future may even feel sorry for us in our day because of the level of misunderstanding that prevails in our time. The second horn just like the other nine horns are a part of a shape or form that was created shaped and sculpted by a shadow and it consists of light that has been changed and made to last by the Ancient Object's powerful forces. This is all a part of the evolving reality of the Lasting Light event. It turns out that many of the passages in the bible that people interpret as having a symbolic meaning are in fact literal descriptions of sights that the ancient writer actually saw when they were looking up through and seeing from up within the region or area of Lasting Light. Attaching symbolic meanings to these literal sights does nothing to accurately describe these things.

The symbolic meaning given to these sights in our day have no true meaning or relevance at all as far as describing the Lasting Light and the many shapes, forms and sights seen within this incredible event. Many of or more like most of today's interpretations as to what it all means turn out to be basically completely inaccurate as far as what really happens is concerned. My return descriptions will stand the test of time.

Each Standing Stone Crater on the Object's surface cast a number of individual shadows and these individual shadows each sculpt their own fantastic shapes that are all made of the Changed Light from the Moon. The initial sculpting of human looking forms is caused by a group of main most important shadows. I think that there is a very good chance that this first and most important group of shadows is very likely cast by a single most important circular crater's vertically standing crater rim pieces. A Standing Stone Circle on the Earth is actually a model of a Great Standing Stone Crater located on the surface of The Object Of The Crossing Down. I just described a very small portion of what actually happens. It just goes to show you a bit about how far apart some of the ideas of today actually really are from some of the exact details of what really happens. Like I keep saying this is a very troubling and also a confusing situation. A situation that I know I have a duty to point out. I have no other option. As a part of my goals related to my Return Viewer's Guide and the obligations I have to everyone, past, present and future, I have decided to say it like it is instead of beating around the bush or worse not saying anything.

I hate to say this but one day many of the guesses of our time will be a part of the laughing stock of the future. I am not laughing now and I won't be laughing then either. Instead I can't help but worry about this situation that we are all in today in our world's modern times. This is to serious a situation and I can't find any humor in any of it. People just don't know what actually happens. The returning Object and the Lasting Light event that the Object causes are simply the facts of life and the way things really are in our area of our solar system. The size of the shock to come is unimaginable to me. I don't laugh now and I'll never be laughing at today's sincere attempts to understand the many ancient clues around us unlike how the unfair arrogant jeering crowd of tomorrow will be laughing at the guesses of today. I don't laugh instead I worry about the unfortunate situation that we the people of today are all in. I can point out this unfortunate situation and I know I have to because I did see the Object and the details of the shapes within the Moon's hanged Light.

I suppose it could be said that if people are going to venture into what happens to be the area that has to do with the returning orbiting Ancient Object and especially the Moon's Changed and Lasting Light then they are faced with the reality that their guesses and ideas will inevitably be put to the test. I'm sure that this is way more than these people bargained for when they got started. They knew they were working on the world's greatest mysteries so it could be said that these people knew ahead of time that at best they would be standing on shaky ground perhaps one day finding themselves without a leg to stand on if it turned out the bottom happened to fall out on their guesses and ideas.

Incredibly that's exactly what's about to happen because it turns out that many of the big areas of expertise that these people claim to have are very often based on the celestial Object that I describe. The Ancient Object that I saw orbits down across between the Earth and the Moon perhaps every forty six years as a part of it's overall orbital pattern or cycle. A confrontation between the Ancient orbiting Object and the hopelessly off guessers of our day is about to happen. Today's guesses stand zero chance to survive and will be crushed once they come face to face with the weight of the Ancient Object of old, Adoil, who knew? Apparently basically no one. I know that the guessers of today only have a few short years left before the ride that they are on ends. Either way if it's someone else's bandwagon they have jumped on or if it's their own guesses at things the ride is ending very soon.
Will the world once again start counting the years starting back at year 01 again once the facts of the returning orbiting Object come to pass

down across in front of the Moon once more and the face of the Man of Light is seen and known again in our time? We will see...

AR1 After Return, year 1, it's got a nice ring to it. However using the 1BC, 1AD naming convention 1AR follows. If the people of the world decide that it's actually time to start counting the years back from year #1 again we will have to figure out and agree on what we are going to call it. I think of AD designation instead as the time before the Return BR for example 2017 BR then 2018 BR. In this example year 1 AR follows with a new date for New Years Eve, maybe May 26th.

This does relate to the size and the gravity of the situation that is about to occur again and the potential possible reasons for resetting the world's calendar back to year one. Either this is so or this celestial event has nothing to do with the reasons why people would consider a need to reset the world calendar back to year one. Maybe people will think that this celestial event, the return orbit of the astronomical object that I describe and the way it changes the Moon's light and the amazing muscular human form that is seen within has nothing to do with the reasons why the world calendar might get set back to year one. A big part of this is about people knowing having seen the Moon's Changed and Lasting Light and then later deciding for themselves whether or not setting the world's calendar back to year number one is relevant, warranted, needed or wanted. Once again the Awesome Good, the Ancient Man of light is going to stand up and turn around and look at all viewers directly face to face. Even if we don't reset the calendar back to year one a new era in the history of humans on Earth will begin.

The Object is not going to slip by unnoticed again like it basically did the last time it crossed down in front of the Moon. Once people know this for themselves everything will change and nothing will be the same ever again. I believe that the next year one is nearly here again. By a complete lucky fluky chance I saw the Object and the face of the Man of Light! By luck only I know what happens. I am not guessing as far as the basic details and facts of a Return event are concerned. I have no concerns or worries whatsoever as far as my descriptions of the Object and the first part of the Lasting Light event as I describe it being put to the impossible test unless you've seen it that's literally fast approaching it's crossing point down between the Earth and the Moon.

Again one of the main points of this chapter is that the guesses of today are completely wrong. It's obvious to me that unless you've seen what actually happens and were it actually happens the details of a Return event are impossible to guess correctly. So far from the guesses and theories that I have come across so far no one is able to describe the Object's exact details or describe the exact details of what the Object does to the Moon's light while it rolls at you from in front of the Moon.

I provide very exact details of the Ancient Object of the Crossing Down not vague wide ranging descriptions that could be taken to be associated with this or that type of similar celestial object imagined and dreamt up by a guesser or an enterprising BSer or actually real having been observed. There is only one Object that crosses down between the Earth and Moon who's shadows does the fantastic specific detailed sculpting of the Moon's Changed and Lasting Light as I saw and as I describe. It's important that people have some sort of clue as to what really happens. No one needs to believe my Return descriptions however the details that I provide for people are very important and need to be known for many obvious reasons.

People read the ancient texts that contain descriptions of the Object and the Changed Light scattered throughout through less than 100% perfect translations and then they come up with interpretations and meanings. It seems that if everyone or if at least some people agree then somehow these interpretations and guesses are then considered to be solid facts even though the actual true facts concerning the returning Object and the Lasting Light event that it causes with the Moon's light are unknown and not a part of the equation of understanding. Because the actual real details and facts concerning a Return are not included in today's ideas it turns out that these guesses and ideas and facts just simply don't come even remotely close to describing what actually really happens or even where it happens.

Like I said before we are all living in the high tech dark ages of today. At least that's how I think of the type of times that today's world is stuck in. The guesses of today are hopelessly off because people today don't know what happens when the Object returns and intercepts and somehow changes the light from the Moon. Nobody lives there but incredibly the human form is eventually seen as shaped lasting Moon light! This is true and this is very difficult to quickly easily describe and impossible for people to be able to imagine as a way for them to be able to have a chance to guess correctly about what happens.
Obviously the ancient people saw the Object and the shapes within the Moon's Changed and Lasting Light and they wrote about these sights

sometimes in a literal sort of way. These writings are found within many of today's religious ancient texts and other ancient texts. These texts are descriptions of what happens when the Object rolls down across in front of the Moon. Also it appears clear that within many of these ancient texts there is a great deal of content that has nothing to do with the Ancient Object or the Changed Light. Perhaps this content was added later by individuals who knew nothing about the Changed Light?

When the returning Object crosses down through between the Earth and the Moon the Object intercepts and somehow effects the light from the Moon capturing it and slowing it down perhaps and makes it appear solid looking. The behavior and appearance of the Moon's light becomes changed. The Object's vertically standing crater rim pieces cast shadows into the light from the Moon. Somehow these shadows sculpt and shape the Changed and Lasting Moon light. Along with many other types of sights the human form is seen not just once but many times over. These facts are at the heart of the today's confusion and total misunderstandings.

The Object returns and when it does it causes a fantastic event to occur! It's a returning object and a fantastic visual event that repeats in exact detail! This is one of the points I'm always' trying to make. Yes the Object is completely spectacular beyond belief but the Changed and Lasting Light event it causes is spectacular literally beyond all measure and comprehension! The Changed and Lasting Light event was always the spectacular thing or event the the ancient world was always focused on and this is without question.

A Return event is the big thing that happens from the sight of the Ancient Object rolling down to it's crossing point in front of the Moon all the way through from the first sight of the Moon's Changed Light and then the shadow dropping and passing over the Object's Standing Stone Craters into the Valley to well well beyond the point were the incredible very muscular Man of Light, the Awesome Good turns around to his left and stands up and seems to notice you going directly face to face with you the viewer.

Obviously the entire ancient world awaited each and every Return event with great anticipation because of the sights that they knew they were going to see within the Changed and suddenly Lasting Light from the Moon. Incredibly somehow when the points of the main group of tall pointy obelisk shaped shadows reaches a certain point in the distance of the ancient Valley the changed and very different behaving and looking light from the Moon these shadow's individual points sculpt and shape forms out of and with this Lasting Light. The hard part or at least a big part of the hard part is the fact that human looking forms are somehow created.

Once again these basic facts are at the heart of the confusion in our time. Nobody lives there but the human form is somehow generated and created there. Again this reality is at the root of modern day confusion and misinterpretations and lack of understanding. Today's overall lack of knowledge has led directly into today's unfortunate overall lack of understanding.

When the point of the big main shadow strikes and draws the squiggly line perhaps on the surface of the Moon suddenly everything that you are looking at changes and all the shadows are gone and the place were you see from isn't racing forward anymore. Suddenly instantly you are looking down from above and behind and there incredibly right in front of you is the human male form. The very startling intense crystal clear sight of the very muscular back of the Awesome Good!

There's no way that I can describe this sight without the listener or reader just naturally presuming that I am talking about a real human up there sitting or crouching looking out into the valley and then gently rising up out of the flat layer of white clouds of the Valley floor twice directly towards you as I describe. Even after my best efforts the listener or reader thinks I'm trying to describe a real person and or also other real people or aliens sometimes all a part of some sort of an actual civilization of some type. It turns out that this is very hard to describe to someone else without them automatically thinking that you are actually talking about real people or a being that is alive the way we humans are alive after all what else could I be talking about and what else is a person to think when they first learn from me about what really happens?

This is the same effect that happens to people today when they read ancient Return descriptions like the descriptions that include the Anunnaki and the Nephilim for example. People just don't know that there's such a thing as shaped and sculpted Lasting Light never mind the fact that these forms could be in the form of human beings.

I try to explain what happens to the point of the big main shadow to people who speak english just like I do. My listener does not have to try to understand the words that I am using when I am trying to explain these ancient sights. People I talk to have the benefit of a first hand eye witness account and they do not have to use or rely on translations or some sort of sketchy deciphering of ancient texts in a difficult to understand and interpret ancient language in order to hear about what happens when the Object returns and intercepts and changes the Moon's light and the shadow strikes.

I saw the Object and I saw how it changed the light from the Moon. I saw how the end of the big main shadow, The Black Road shapes and sculpts the Moon's Changed Light and joins with it and creates and becomes the form of the Awesome Good. In English the only language I know I try my best to describe what happens and what I saw. In spite of this people are not able to grasp the idea of what I am describing.

Obviously one of the points that I'm trying to make is that unless you have seen the Object and the Moon's Changed Light for yourself it is very hard or nearly impossible to realize from anyone else's descriptions of a Return ancient or modern what is really being described and talked about even if it's in your own language.

I certainly can't explain this event but I can describe what the sights look like at least during the first part of the event. I didn't see the entire event. The Lasting Light event was still progressing and evolving and growing and the Object was still visible not having finished rolling away yet in the lower right area of the view finder when I stopped looking and my view was over.

I did see the muscular human shaped form of not just the Awesome Good but I also saw the other incredible human forms that the points of the rest of the main shadows created in the distance of the ancient Valley of Light. The light from the Moon becomes slowed down and perhaps even stopped and then it lasts. The human forms and shapes that were sculpted and shaped by the shadows are made of light. The Moon's light becomes effected by the Object and somehow it lasts. The human forms that you see are actually Moon light that is solid looking as if statues. The impossible is there and it has human form. Not only that but the impossible repeats itself many times. I have to try to be clear here because this is obviously very difficult to explain clearly.

The heart of the Lasting Light is the incredible Awesome Good, The Man of Light as I try to describe. The impossible that repeats is in the human form as well however these other forms that are suddenly seen are different human looking forms compared to the Awesome Good and they all happen later after the Awesome Good was created.

As a kid and growing up and still to this day to some degree I always have and always will think of them all as, The Guy and The Crowd.

Down within the Lasting Light Mountain's upper structure at the top of the rising column of lasting Moon light is the sunken inner place were the Guy and The Crowd are located in their curving staggered semi circular pattern right after the crowd is newly formed and created by their own individual shadows. Shadows that were each cast by a vertically standing crater rim piece. Again it could be true that all of the main shadows were cast by the standing rim pieces of a single crater. I happen to think that there is a very good chance that this single most important crater with it's vertically standing rim pieces is located just beyond the curving top of the Object's raised surface feature, the mound that eventually becomes the Overhang as the Object rolls forward, a guess.

The column or tunnel of rising Lasting Light with the sight of the Awesome Good and his accompanying friends is transported straight towards the Earth. That is what was happening when my view of the Object's last Return ended. In very ancient times I believe that the area down within the rising upper structure somehow continued towards the Earth while the sleeve or tunnel of light that transported this layered internal Lasting Light down within becomes left behind. This is very complicated and I am going to be continuing to try to describe the sequence of events while mentioning and describing the evolutions of the various separate elements that make up the Lasting Light event as far as I know it and can guess about it on my web site.

Today people have no idea that the situation and the scene as I describe it is actually what really happens. Today people translate and interpret various ancient texts and based on these translations and interpretations and based on what other people think they try to figure out what it all means. I see some people being very careful not to deviate very much at all from what those they respect think so that they will not find themselves being ridiculed by these people.

I see some other people looking at various different sometimes obscure ancient sources and then saying it like it appears to them without regard for what anyone else thinks and completely unafraid of anyone's ridicule or condemnation. These are the people who's work I find most interesting. It's just unfortunate that they don't know about the Moon's Changed and Lasting Light and the crossing Object. If they knew about this then their collective understandings would be nearly complete because this is what actually goes with all the fantastic ancient facts that these people know first hand for themselves as a result of many years of dedicated hard work research and study.

If the ancient material that people today are dealing with has nothing to do with the Object and the Changed and Lasting Light from the Moon then I don't know anything about it and I can't comment on any of it. I also have to say that if the material being dealt with is about the Object and the Moon's Changed and Lasting Light then that's what it is and that is what it's all about. People don't know this and the fact that they don't know this isn't their fault or anyone's fault.

That said it does not change the plain and simple fact that if the ancient texts or depictions are about the Object and the Moon's Changed and Lasting Light then that's what it is and that's what it's about and definitely nobody lives there on the ancient Object of the Crossing Down and that's the way it is.

9 THE SHAPE EFFECTS

Introduction
Writing about or talking about the effect that I try to describe as the shape effects is not easy for me to do in a way that makes any sense to anyone. At least that's the bottom line reality for the normal person like myself for instance. If I didn't see and know this for myself there's no way I would be able to consider the Return Viewer's Guide and especially anything to do with the Changed Light as possibly being real. Like everyone else I would have to say to myself that none of this could be true because if it was true everyone would already know about it and since we don't know about it it simply can't be true. That's how I would feel about if I didn't know about it and I happened to stumble into the Return Viewer's Guide. In today's world there are lots of ideas and theories out there. They range from being really interesting and credible to far less interesting and credible. Certainly in the broad spectrum of things the Return Viewer's Guide and the Changed Light and especially the shape effects would and could only be placed at the bottom by everyone including me. As it turns out today this situation has many effects on me as I try to do the right thing. There is a type of panic because I know I'm alone with this problem. This is way to big for me alone but I am alone. If I could take this whole thing and condense it down into an object I could hold in my hands instantly I wouldn't know which way to run or what to do with it all the while I would be terrified that I would drop it! This is to big of a thing for just one person. This is to big of a thing for everyone else as well because it's an individual thing. It's just you with suddenly nothing in between you and the Object and the Man of Light except for empty space. It's just you and him and the

terrifying Object. As soon as you see it you are stuck with it with no one to help you. For me later and today my heart skips a beat because I know that what I saw is primordial. It's an overwhelming thing for me because I know that there isn't any chance that there is any example of the formation of the human form that precedes the creation of the muscular Man of Light! This stark reality makes this whole thing different than any other type of problem that I could imagine being faced with. The tremendous gravity of the situation that the world is in today with the Object nearing and the tremendous responsibility that I know I have to at least get part of this right so that I can make a difference and help everyone including you has always crushed in on me especially since my children were born. I know only a small fraction of how important the formation of the Man of Light is and I know that there is a good chance that the effect that I try to describe as the shape effects is important and in spite of how strange this effect is and how impossible it should be and how it is unlike anything else that I know of and how hard it is to adequately describe I have to try. Anything that has to do with the answer to the question why are we exact copies of the Man of Light is important. This is why I know I have to try to explain the shape effects. When I wrote this chapter I found it hard to do so in logical well structured way. Actually you could say that I have the same problem with everything I write.

Patience is will be required, thank you.

The Shape Effects

The first time I tried to describe what I had seen was seconds right after my view of the incredible crossing Object. I was a kid and I turned to my friend's Dad who had also just seen. He only saw for a split second before it was over for him but he did see. Totally stunned and shocked and grimacing with his teeth tightly clenched Herb looked right at me his eyes wide open! Immediately I started to describe the sights from the beginning of what I had just seen starting with the sight of Object itself during the time before the Moon's Changed Light had happened, the actual forward rolling crater covered massive boulder itself.

"NO!!!" he shouted me down. It was not his fault. It was many years later before I suddenly finally realized that he knew nothing of the Object's hard rolling ground. He had no chance to know about the Object itself. I had wanted him to see to see the Object before it left the area within the telescope's viewfinder and that was one factor and one of the reasons why I stopped looking when my view actually ended. For a split second after my view was over, for a split second he had seen. By then the forward rolling Object was well further down in the lower right area of the view finder if it was still actually visible at all when he looked. I think it probably was but I actually don't know if the Object itself was still actually visible to him. I clearly understand that during the split second that he saw for he did not see or notice the actual crossing forward rolling speeding Object itself.

Instead he saw the heart of phase two of the Changed Light event, the rising Changed Light Mountain. He saw the second evolution of the ancient Man of Light, the Awesome Good! Not only that but he saw him at a closer position compared to the point he had reached when my view was over. It's obvious to me that the intensity and pressure that's simply off the scale was there and he had suddenly felt it full force! If you don't see everything from the very beginning clearly you can join in. He started to look down right away after I moved away. He never made it all the way down to the eye piece before he grabbed the telescope to his eye and stood up all in one fast motion and then looked right at me! The Moon's Changed Light had reached right up out of the telescope all the way to his eye well before his eye had reached the eyepiece. When that happened he was instantly struck with the intensity and pressure that is so intense and incredible that it is simply beyond belief. Seeing is what it is all about. I know for certain that without any question Herb did see and feel the pressure from the Changed Light from the Moon!

I know now that as he suddenly stood strait up he was well past complete shock! At the time I did not understand what had just happened to Herb. When he shouted "no," at me I thought that he was arguing with me and I suppose in some ways he was arguing or disagreeing. At the time I simply assumed that he knew all about the big place and what happens there and everything about what I had just seen. Now I know that he was struck by the pressure and the sight of the stare of the Man of Light at the heart of the rising Changed Light Mountain. That's what he saw and that's why he had no idea about the celestial object I was attempting to describe to him. I now know that he was completely taken by surprise and he was just as unprepared as I was at the time. He went from not seeing to receiving the full force and intense pressure of the stare in a split second for a split second. I can but at the same time I can't imagine how difficult that must have been for him.

At the time I was no were close to understanding these circumstances instead I literally felt as though I had somehow made him mad at me because of my response to him when he glared at me after he had stood up straight after bending down to look in the telescope. After this totally disconnected exchange with my friend Mike's dad he literally turned and walked strait into his house with the telescope that he was already holding in his arms with my my smiling and confused I suppose friend Mike in tow.

Badly shocked and confused and very scared frankly by everything that had just happened I hurried home struggling to see past the glistening white shape effect that was just there in front of me as I walked home in the dark. I had no idea the white glistening shape I was seeing out in front of me was unusual and not normal and although I was thinking about it I basically ignored it as walked south up the hill between our houses crossing Birchview Dr. towards the end of my driveway towards my house on the other side of the road.

To scared to look back towards Mike's house because I was certain that something probably really bad was going to suddenly be happening in the exact spot where the telescope had been set up so instead through my fear I looked up into the sky to my upper left and except for the Moon nothing was happening there which was a tremendous relief but this also served to cause me to become even more confused. How could there just be nothing up there right where I had just seen all this incredible stuff going on? I couldn't hear anything happening at Mike's house which made no sense to me either To scared to look back and with a stone block embankment wall blocking my view of Mike's house my imagination ran wild as I was picturing everything and everyone, all the human shapes that I had just seen, everything accept for the Object somehow in Mike's yard! Right up in the sky where there was now nothing happening the Man of Light had been right there with his friend's literally seconds before this and instead under only the Moon and an empty starry sky there I was quickly walking home on a empty street in a quiet peaceful neighborhood. Nothing was making sense to me and this didn't make any sense either. Where were the air raid sirens that I had heard before for what seemed like no reason? Why wasn't I hearing the sirens? Where were all the people? Surly everyone would be coming out of their houses and also looking up to see what was happening. Eventually I looked back once and nothing and up in the sky there was only the Moon! How could this be? Where was the big thing that was bearing down on me just moments before? Suddenly a small sense of relief started to creep in for the first time. Everything seemed normal but how could this be? Nothing seemed to be happening. As badly confused as I was I only grew more confused all the time. It was past dark but still early evening and I was home early but I know I went strait to bed. Then suddenly in my dark room the large shape effect was just out in front of me where ever I looked even more pronounced and noticeable than even before and I think this is because I was in the darkness of my room.

I was thinking that this very strange very pronounced sort of afterglow effect was just a normal thing that happens after looking at something bright through a telescope. I had never looked through a telescope before. I remember thinking that this strange effect must be happening in this unusual way because the Object was so bright. I did not give it much thought at first but with this glowing effect so hard to ignore I remember that I was starting to look at the shape effect's sparkly crystal like edge. I know that I was starting to really wonder what I was seeing because it was a weird thing or effect that seemed to be like a cloud that was happening all by itself a real thing in front of me wherever I looked out from the wall towards me. When I looked around it also moved so I knew that the shape I was seeing had something to do with me but if I just looked at it the connected sensation subsided and instead the large shape effect seemed like it was over there all by itself taking up space in existence, and some sort of real thing. Eventually I feel asleep.

This cloudy effect that I am trying to describe is the effect I call the Large Shape Effect. Take your right hand and hold it out in front of yourself. Spread your thumb and fingers out wide. Rotate your hand counter clockwise to the left until your thumb is pointed down. This is the overall shape of the entire area of the Changed Light during the first part of the Change Light event when the Moon's light initially emerges to the right of the Object and looks like a valley.

THE SHAPE EFFECTS

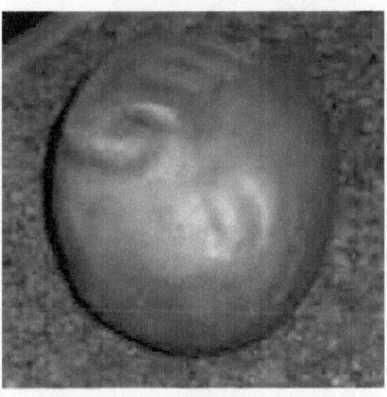

This clay model of the Object features a simulated imprint of my right hand on the right side The outer perimeter area between your downward pointed thumb and your forefinger is the upper right horizon

I think that the finger effect that is a part of the large shape effect is caused by the stark contrast between the Changed Light and the long tall black individual shadows that follow to the left and possibly to the right of the main central shadows. Or the shadows that cause the lines or finger effects within the large shape effect's area are caused by just the shadows to the left of the three main central shadows. I believe that it is more likely that these shadows on the left are responsible. At first these shadows look like fence posts and then they look like telephone poles as the many individual shadows lengthen and grow. Because I know that the vertically standing crater rim pieces on the Object are casting these shadows I am caused to conclude that this group of shadows might be cast from the same crater because of their uniform look.

I am convinced that the force that causes the large shape effect is the same basic cause or force from the Changed Light that caused us human beings to come into existence. I believe that the shape of the outer edges of this cloudy shape effect matches the shape of the overall perimeter of the entire Changed Light event during the first stage of the event when you see a tremendous valley to the right of the Object and over it's upper right horizon. I believe that somehow the Changed Light has the ability to imprint itself into the viewer whoever that may be. Somehow we are copies of the muscular Man of Light. It's my guess that somehow some of the life that saw the Man of Light was caused to grow and copy and shape itself and eventually evolve into an exact copy of the man of Light in extreme muscular detail. The pattern is displayed within the Changed Light and we are the copies of that shape. It's just way to bizarre but it appears as though at a basic level this does seem to be the case, this is what happens.

Over the next few days after I saw the Object a very small sense of relief started to grow. Slowly it became more obvious that nothing had happened or seemed like it was going to actually happen and my sense of relief gradually began to over power my fear and confusion. As the immediate afterglow and the strong freshness of the sight of the large shape effect faded I can only suppose that the areas of my mind that were effected by the Moon's intense Changed Light also slowly began to somehow gradually simply shut down. Not completely but I am fairly certain that there was some sort of a shutting down effect that slowly happened to me over the next few weeks.

I describe this situation in the Return Viewer's Guide and on my website in various places. I believe that the total effect of all the circumstances combined with everything I saw and everything that happened to me plus being taken completely by surprise thinking I was going to be looking at the Moon again on my next turn looking through the telescope, was to cause the Changed Light's effects to magnify the wonder, shock and the fear that I felt. I honestly feel that if you have a good experience then that good experience is what becomes magnified and this becomes how you remember and how you instantly strongly feel later when you are remembering and thinking about it in a strong way. My experience was true wonder, true surprise, true shock and true fear. I know for a fact that a certain minimum level of fear is part of seeing starting with the very intimidating Object and especially so concerning the Changed Light especially when it's rising and you can feel the pressure and you are looking face to face with the rapidly nearing Man of Light. It's very complicated. The shape effects don't just allow you to see the cloudy white after glow effect again many years later but also how you felt while you were seeing is brought back to you at the same instant and is magnified inside you and causes how you felt while seeing to happen again in a way that is very fresh and real. In my situation I knew extreme shock and fear. I feel very strongly that the viewer is suppose to have the opposite experience and I believe that a properly prepared viewer can achieve this very worthwhile goal. My advice is to love him back as hard as you can and joyously experience every split second of your time seeing from within the Changed Light. Later this will be how you are instantly made to feel when you are concentrating hard and seeing the afterglow again thanks to the shape effects. I think that the ancient people would have taken this basic concept to a highly advanced refined level of practice. I believe that this may be just the tip of the iceberg.

Although I am very lucky and very very fortunate to see the Object and the Moon's Changed Light it is unfortunate that I didn't live in world and we don't live in a world where everyone knows about the Object and what happens and especially a world where people understood the Changed Light and it's mysterious effects and power at least to whatever furthest point possible. I hope that one day in the future the children will be taught everything that they need to know to be able to see the Changed Light with not just foreknowledge the way I am able to offer people a basic road map but my hope is that in the future the children will be able to see for their first time from deep within the Lasting Light with true understanding and foreknowledge thanks to their parents first and thanks to the collective efforts of everyone in the future who tries to understand what is happening to the Moon's light and what happens to you when you look at this light. Knowing ahead of time that this is very profoundly special and to concentrate on every second of time and every sight as you see is important. I believe that the shape effect's causes and effects and imprinting effect on a human who is looking and concentrating on everything and truly enjoying and loving every element and everything they see in every aspect will have their experience and memories enhanced.

Because I knew so little or rather nothing at that time I did not know that the age old human tradition was to now focus and concentrate as strongly as possible on the cloudy afterglow effect that you can see immediately after seeing the Object and the Changed Light. After reading about some ancient traditions that do certainly sound strange at first and based on my very limited experience seeing the afterglow large shape effect again 35 years later after seeing the Changed Light, I believe that in fact traditionally the point is to remain intensely focused on looking at the afterglow effect that you see happening right after seeing the Changed Light while also thinking about the things that you just saw happen. If you had a wondrous experience seeing starting with going face to face with the Man of Light and throughout every aspect of the entire event then I'm certain that this could only make it easier for the person thinking about this to be able to experience and see the shape effects later because it would be such a welcome feeling. Over time if you continue to regularly concentrate and focus on what you saw I feel as though you could experience seeing the afterglow again and feeling the strong remembrance of the emotions that come back to you from this special past experience for the rest of your life. That's not what happened to me because of the circumstances I was in before and after my view of the Object and the Changed Light.

To be clear I am guessing about certain things to some degree here because I do have limited experience having only seen the large afterglow shape effect right away after my view of this event was over and then again 35 years later. On three consecutive nights I did see the large shape effect again. The third night was the night that everything came together for me and that was the night that I realized that I was seeing the large shape effect and also it was at that instant that I realized that I had also seen that shape in the two previous nights but I was not fully aware of it or more like it wasn't as clear and obvious to me. I had no idea that I was going to actually see the large shape effect again. This was not something I knew about until it was suddenly simply happening and that's when I became aware that you could suddenly be seeing this bizarre afterglow effect many years later. I suddenly discovered the overall shape of things meaning the overall shape of the Changed Light because the large shape effect is the shape of the first stage of the Changed Light event when the sight of the ancient valley is seem. When I found myself looking at the large shape effect I could see the Object's upper right horizon within the large shape effect and this allowed me to see and realize the shape of things again. This mattered to me for many reason and one of then was it was a fresh reminder or fresh look at the entire situation and specifically the border or boundary between the hard rolling surface of the Object itself and the Changed and Lasting Light and predominantly this border area is the Object's upper right horizon and also the right side of the Object.

When you look at the Moon's Changed Light to the right and upper right of the Object you see what looks like a valley that is surrounded by mountains on three sides. This scene is oriented for the viewer exactly the same way you would see a valley with mountain ranges on the Earth. The picture is right side up and you are amazed by this fantastic scene that spreads out extremely wide and deep and it really does look like a real place. There is no way that you are going to see this and realize that it's the light from the Moon. Understanding that everything that you see that is not the Object is actually light from the Moon is not going to be something that is easily realized. For me later when I saw the large shape effect again it allowed me to see what was Changed Light and where it's edge touched the edge of the Object's upper right horizon. Seeing the large shape effect again allowed me to see the exact way that the Changed Light is shaped looking like a right hand as I described above in this chapter.

Again this comes back to the ancient mystery at the heart of what's important. We are an exact conscious living matter copy of a shape that is made out of the sculpted and shaped solid looking Lasting Light from the Moon. It's just to bizarre! What has to be a randomly standing vertical crater rim piece on the crossing Object's surface casts a shadow into the Moon's Changed Light and that shadow does the I can only suppose random things that it does when it interacts with the Changed Light and it creates the muscular human shape out of that light and we are an exact copy of that shape. This has always been a shocking thing for me along with knowing that this is directly connected with everyone around me and yet not a single person around me is aware of this situation and the facts as they really exist. For me this all contributes to the magnitude of the shock to the world that is going to happen when the world sees the Man of Light again and realizes these things are connected to so much of our world and our different cultures and ourselves. There are so any things that are just to bizarre about this situation and then for the reader hearing me talk about the large shape effect they have to be left even more confused about what I am trying to talk about. I am certain that the process that caused us humans to be copies of the muscular Man of Light has the large shape effect as an important part of that process. I describe the large shape effect afterglow exactly the way it appears to you when you see it. It's like a cloud in the room between you and the wall. Describing this effect is important and although it makes me sound a bit nutty, and that's being kind, I have to do what's right and if it's important and I know that this is important then I have to talk about it going on and on talking about it until I have gone around in circle a few times.

The process and the energy and intensity from the Changed Light that causes the light's ability to imprint itself into your memories is at work right from the first instant of the valley when it's a tranquil fantastic peaceful scene. The shape of the valley's outer perimeter at this point matches the shape of the large shape effect. Also again this is the shape of the afterglow effect the viewer will see out in front of them at a short distance everywhere they look after looking at the Changed Light. You could loosely compare this effect to when you look at a bright light in the dark and then right after the light is turned off in the dark you will be able to see the glowing shape of the light for a short time. This example does illustrate a similar afterglow effect that we are all familiar with. At a biological level there is probably a direct connect between the light bulb afterglow and the afterglow from the Moon's Changed Light.

I saw many fantastic things and I know what happens and it's overwhelming. The things I saw just shouldn't be able to happen. The stunning powerful Object and the intense Changed Light and a world around me that is obviously unaware causes me a level of stress that never goes away. I have always struggled greatly not just being scared of the Object and greatly impressed by the Last Light sights but I believe that the shape effects causes me to have a particular strong reaction to being suddenly taken by surprise when something unexpected reminded me about the Changed Light. This definitely effected me a great deal more when I was still young however this effect still happens to me automatically. If I start to tell a person about this then instantly before I do actually say anything my heart rate spikes and I forget to breath when I talk and I start to sweat. I know that the physical reaction that I have is tied together with all of the effects that seeing that Changed Light has on you.

In order for me to see the large shape effect again I had to force myself to concentrate hard on my memories of the Changed Light. In hindsight I now realize that I somehow knew that I should do this and I was doing this already because there was something there that was very familiar and real but I didn't understand anything about what was going to actually happen when I suddenly found myself looking at the large shape effect again. Why am I telling you about this and why do I think that this is important? Somehow some of the life that looked up and saw the muscular Man of Light was caused to grow into and copy the very shape that it was looking at and this is a mystery or I suppose that you could say that this is the mystery at the heart of who we humans are today. The afterglow effect that happens to you automatically as a result of looking at the Changed Light is connected to the imprinting effect that the Changed Light has on the viewer. Ultimately this shape effect that you can see is connected with the mystery of how and why are we exact copies of a shape that is light? It is important for future viewers to be as prepared as possible and knowing about the imprinting effects that the Changed Light has on you and that this is going to happen to you is something that I believe people would want to know.

During the weeks and days that led up to when I saw the large shape effect again in approximately 2003/4 I was concentrating on my memories of the Object and the Changed Light. I was remembering what the overhang looked like were it joined with the Changed Light from the valley. I was trying to recall how far I could remember what the overhang looked like for as far as possible to the right towards it's joining point with the valley. This position or place is located to the far left along the line of the Object's upper right horizon. As it turns out concentrating on this location trying to see further to the right towards the place in the large shape effect were the solid ground of the Object ends and where the formerly eclipsed Moon has emerged caused or triggered something for me because this is exactly what I was doing and what I was thinking about when I found myself suddenly very expectantly actually seeing the afterglow large shape effect again. This place is the boundary between the light from the Object and the Changed Light from the Moon. It's a blurry boundary that looks very similar to the rippling one way moving areas of Changed Light that flows from the upper right horizon to join in and become new base area of the rising Changed Light Mountain. In both of these examples of the light acting this way the light is very close to the surface of the Object. The blurry edge as I refer to it is not actually blurry instead it's more like a rippling brook or stream in the sunshine with light reflecting and glistening and flowing.

The farthest point to the right along the horizon edge of the Object's mound or overhand where it intersects with the upper right horizon's glistening rippling edge were everything smoothly transitions into the Moon's Changed Light is exactly were I was looking and the spot were I was concentrating when I suddenly realized that I was actually seeing the large shape effect. During the preceding two nights I was close to seeing the large shape effect but it didn't happen the way it happened on the third night when I fully realized that I was seeing the large shape effect. If I had known ahead of time that this was something that could happen and was suppose to happen it would have been much easier to actually make or have this happen. On the third night wide awake and fully aware I saw the large shape effect again less intense but just as real as the first time I saw this shape right after seeing the Object on my way home and later in my dark bedroom that night while trying to fall asleep.

I am taking my best guess concerning many things and this is another. I am guessing that if you love him back and you knew joy and love when you saw him then that's exactly the feeling that you would suddenly feel and have strongly within you when you really concentrated and focused on these memories years later. I am basing this guess on my past experiences over the years when I start to think about and sometimes talk about what I saw. I know that I was delayed in my efforts to try to think about and understand and describe the Object and the Changed Light. I believe that this was the case for many reasons but one main reason why I believe I had difficulties was because of the fact that I was unprepared when I saw the Man of Light and I experienced true surprise shock and fear and these are the emotions that rush back to grip me instead of for example joy and love and happiness. I believe that in combination with whatever happens to you from seeing the Changed Light that causes the large shape effect to be seen again later even years later, I think this effect is also responsible for the way memories of how you felt while seeing the Man of Light are remembered in a very immediate fresh way.

This is something that may be important in the future for people who will be studying this event. Do my observations actually point in the general direction of understanding? I think so but I really don't know but in spite of not knowing the answer to this question as a part of what I feel is a responsibility that I have to try my best to report, explain and describe what I saw I have to also talk about the very strange shape effects and how I have noticed it's effects on me even though it is really not something that I am very comfortable talking about. I certainly am not worried about whether or not people think I'm a bit off my rocker and who knows maybe I am but in the meantime I did see the Object and the Changed Light and I have to try to do what is right. Without any question whatsoever something does happen to a viewer who sees the Changed Light. It's important that people have a chance to hear about this effect so that along with being better prepared to see the Changed Light they also will be in a position where they will be able to look forward to seeing the special Changed Light with excitement and anticipation knowing that it's actually good for them and as humans they are suppose to see and look at the Changed Light.

As an unsuspecting 12 year old boy in 1972 I was not equipped be to able to begin to understand even the most basic aspects of what I was seeing. Jumping ahead to the heart of the matter that 12 year old boy had no idea that the Man of Light wasn't an actual real guy. I had no idea that he was made of light. I literally took me the rest of my childhood to begin to understand that he wasn't really mad at me for disturbing and looking at him which sounds strange but that's how it was for me at that time. After watching the shadow streak into the distance of the valley and then seeing it strike and draw the squiggly line I was instantly suddenly looking at this muscular guy who wasn't wearing a shirt. So as I am looking down at him from above and behind suddenly taking completely by surprise he suddenly gets up and turns around to his left all in one fast motion to look directly at me eye to eye as though he knew that I was there looking at him before he got up and turned around. This was a huge shock for me because it really is as if he knew I was there and he noticed me. Suddenly there was this very intense guy looking back directly at me through the telescope from outer space! Again I really did think that he was extremely annoyed and mad at me basically because I looked at him. That was my rational or instant reaction at the time basically that was a thought I had instantly when this happened. I know that sounds bizarre and it is bizarre and that's part of the point I' trying to make.

By by saying and describing what I saw I'm describing a shape that is light that happens the same way and is created the same way every time the big main shadow interacts with the Changed Light. Everything that is seen repeats and it's all very fantastically bizarre. For any person and every person who will see the Man of Light in the future some basic important ideas are that firstly seeing the Changed Light is good for you and you are suppose to see his face because you are human and this is what is suppose to happen to you at least once in your life. A basic point that I'm trying to make is that future viewers and especially hopefully as many children as possible need to have heard the idea of loving him back. When he stands up and and turns around and looks at you love him back! A good starting point for all future viewers is that he wants you to look at him and he loves you and the more you look at him and the longer that you are able to look at him for the happier he is!

The mysterious shape effects produced by the Moon's incredible and fantastic Changed Light will allow you to somehow actually still see the afterglow again years later! I wish I knew this before I saw the Object and the Changed Light. If I had been a prepared viewer and I had known about part of the point of the shape effects then I am sure that I would have felt a great deal better about the whole thing and the entire experience. I feel that this would have resulted in a situation where I would not have had the types of difficulties that I had especially while I was young. If future viewers are able to benefit from my experiences and my comments regarding those experiences that would be positive and important for those viewers as they continue through their lives remembering back to the experience of seeing the Object and the Changed Light and the very very special primordial human sights seen within.

10 THE RETURN VIEWER'S GUIDE

Soon The Ancient Object Of The Crossing Down will return home to our solar system. Shortly after that the orbiting Moon coloured speeding forward rolling Ancient celestial Object will be seen fantastically changing and then casting shadows into the Moon's light as it rolls down across between the Moon and the Earth. A fantastic time in the history of all the people of the Earth is about to take place again!

Familiarize yourself with the details and the sequence of events that I describe. Numbered 1-5 combined these five separate Return Viewer Guides describe what a viewer sees actually happening when the speeding forward rolling Ancient Celestial Object of the Crossing Down Between the Earth and the Moon returns and crosses down in front of the Moon.

Look at and follow the point of the biggest widest tallest obelisk shaped very black shadow as it races into the distance to the right. If you do this and you are able to continue looking you will see and you will know exactly what happens when the big main shadow's point suddenly sculpts and shapes the Moon's incredible changed light.

I am trying to put future viewers in a position were they have detailed foreknowledge concerning the sights that are about to once again happen between the Earth and the Moon.

As long as you are eventually looking to the right at the point of the big main shadow by the time the wave of the shadow is old and the building out that happens is well underway everything that you see and everything that happens just seems to happen naturally. Stay looking don't look away! Looking at the Changed Light causes you to see from the place you are looking at as if you are seeing out of a remote camera up there. If you look away you will have to start all over again. You will see less and this means you will know less.

All five of the above Return Viewers Guides are taken from this page: www.returnviewersguide.ca/returnViewersGuideForUseWithASmallBackyardTelescope.html A printed copy of this free file saved somewhere guarantees you that no matter what is going on at the time of the Ancient Object's return you will have accurate basic viewing information that you can depend on and trust in.

If ahead of time you decided to read even just a small amount for example you read in point form Return Viewer's Guide #1. You would know the basics of what happens to the point where you would know enough about what's happening and what you are going to see so that you could help other people to see. Just quickly tell them the basics as I describe and emphasize the fundamental point that they should look to the right and everywhere but very quickly find and follow the point of the big main shadow as it rapidly travels into the extreme distance of the Moon's Changed Light to the right. Then stay looking and don't look away. Surprisingly this is not an easy thing to do. A person really has to brace themselves for many reasons. To start with it will be instantly clear to the viewer that they are seeing the Big Ancient Thing that happens. The Ancient Object and the Changed and Lasting Light from the Moon. Suddenly the viewer is also confronted with realizing that they really actually are seeing in the ancient way the Moon's Changed Light causes you to see in as was written about long ago. This is a shocking thing to be sure.

This will happen to you if you are looking towards the Moon when the speeding forward rolling Moon coloured Ancient Object crosses down. Very quickly the difficultly level is off the scale. Bracing yourself and forcing yourself to keep looking and feeling the powerful flow of pressure and intensity that is the way the Lasting Light feels is in my opinion what is suppose to happen to you as nature intended. This is a part of a greater reality that exists for everyone to see including you.

At a very basic level I believe that seeing the Moon's Changed and Lasting Light is good for you. From the beginning in those first seconds when you are looking to the right at the Ancient Valley the entire shape of the Changed Light is imprinting into you even though you haven't started to feel the way you will actually suddenly start to feel dramatic force and pressure the way you do when you are looking at the rising Light Mountain. The Ancient Valley is a tranquil expansive place compared to the overwhelming screaming intensity of the face to face rising Light Mountain. Later when you are really remembering this the way the Shape Effect's allows you to you'll see that this imprinting effect is a real and fantastic thing that looking at the Changed Light does to you. This effect is a central part of this entire very special very ancient fantastic natural visual spectacle. Right after the Light's inward pulsing transition time period happens you are suddenly seeing the rising Light Mountain and you are shocked by what you are seeing. Also now you are feeling and struggling with the indescribable sudden pressure and you are realizing that now you find yourself feeling the shaking inside, it's ok and this is suppose to be happening this way and it's not hurting you. If you remember this I know this could help you to remain looking going face to face with the Awesome Good, the incredible Ancient Man of Light!

Return Viewer's Guide #1

1. The forward rolling, moon coloured object, rolls down from the area of space above and to the left of the Moon.

2. The object covers the Moon's upper left area first before it completely obscures the Moon.

3. As the object rolls forward a large raised surface feature comes into view from over the top of the object.

4. The Moon is suddenly seen to the right of the object.

5. The Moon no longer looks like the Moon at this point, instead the Moon looks like a fantastic tremendous valley.

6. As the object continues to roll forward the raised area becomes a smooth curving steep cliff.

7. The raised area then gradually develops into a very large smooth and rounded overhang.

8. Suddenly a shadow that looks like a big black spot appears under the overhang.

9. The black spot of the shadow spreads out and thickens under the continuously growing developing overhang.

10. Suddenly a certain critical point is reached and the shadow suddenly leaps down to the ground below.

11. Immediately the wave of the shadow races to the right and travels away from the viewer at an ever increasing angle.

12. The shadow is going to pass over and eventually travel off the object's surface to the upper right into the Moon's light.

13. A series or group of tall narrow individual shadows emerge and travel further and faster than the shadow's main wave.

14. At one point this series of tall shadows look a lot like a line of telephone poles stretched out across the countryside.

15. The group of tall individual shadows that are seen are spread fairly evenly from left to right beyond the shadow's wave.

16. A group of three main obelisk shaped shadows emerge and travel further, faster than the rest of the main shadows.

17. In between the two outer smaller obelisk shaped shadows, a central larger wider obelisk shaped shadow is seen.

18. The big main obelisk shaped shadow is twice as wide as the two smaller obelisk shadows that trail, one on either side.

19. This central obelisk shaped shadow is the big main shadow that ultimately sculpts the form of the Awesome Good!

20. The central and most important main idea is this; Find and focus on the point of the big main shadow as it races away.

21. To the left, if you see normal surface area past and over the raised area, you have only seconds left! Look back right!

22. Once again, this is important! To the left, meaning, if you are looking at the hard ground of the object itself.

23. To the left, if you can see beyond and over the raised area and you can see normal surface area, look to the right!

24. Look back to the right, right away because there are only seconds left before it's time to get noticed!

25. To the right, locate the point of the big main black shadow, and start to focus on the idea of not looking away!

26. Find the big main shadow positioned centrally between two smaller shadows that accompany but trail on either side.

27. The main now very tall obelisk shaped shadow is now traveling very rapidly into the distance of the Moon's light.

28. Suddenly the big main black shadow's point starts to go up and down and up and down very rapidly.

29. The ultimate moment is about to arrive. You are about to see the sight of the Awesome Good from above and behind.

30. Instantly after the up and down and up and down stops, the shadow's point starts to cut to the left and to the right.

31. The shadow's sudden zig zagging squiggly line is now sculpting, and forming, and creating the Awesome Good!

32. Suddenly the race forward has stopped and the big main obelisk shaped shadow is gone.

33. Now not moving forward or backwards you look down on a peaceful and tranquil fantastic scene!

34. Suddenly, in extreme detail you clearly see the Awesome Good's very muscular back!

35. The view down from above and behind lasts just long enough for you to be able to realize what you are seeing.

36. The sight of the Awesome Good gently drifts up closer towards were you will find yourself seeing down from.

37. Also, a flat area is suddenly seen around were the Awesome Good is positioned.

38. Past the outer edges of the flat area the misty white indiscernible area is still visible. I try to discuss this in detail.

39. Each successive newly arriving one way wave pulse of rising lasting light, builds out, then down.

40. Another instant later, it happens! What must surely be one of the all time most incredible sights in history!

41. The incredible Awesome Good, turns around to it's left and gets up to squarely face you all in one very fast motion!

42. The Big Noticing! The sight of what should be impossible has just spun around and looked you strait in the eye!

43. This ultimate very up close first moment lasts only long enough to go face to face for only a very few seconds!

44. In that time he drifts up twice to his closest point, then seeing from farther back you remain face to face while rising.

45. The first part or phase of the Changed and Lasting Light event, the Valley is about to end.

46. Between phase one, the Valley and phase two, the Light Mountain, a short transition time period exists.

47. The ultimate staring contest lasts from the big noticing through the transition period, into the rising Mountain phase.

48. The transition period between the valley phase and the mountain phase, only lasts for a few seconds.

49. The outer edges or outer perimeter of the Moon's light, everything that was the valley shrinks, or focuses inwards.

50. This is the same point that the position that you are seeing from rushes back away from the Moon's changed light.

51. During the transition a shrinking pulsing inward effect is repeated and briefly seen, a number of times.

52. At this point the Awesome Good wiggles around, reorienting itself moving very much like a cat about to pounce.

53. This is also when the way the Awesome Good looks or appears overall changes and evolves into it's second look.

54. This is also the point when the rest of the shadows that trailed the main shadow are now sculpting their own figures.

55. This is also the point when the Moon's light seems as if it is finally arriving from over the object's upper right horizon.

56. Instantly after the inward pulsing effect is over the full force of the pressure and intensity of the Moon's light is felt!

57. Simultaneously an incredibly startling sight is seen as the entire area of the Moon's now solid lasting light, rises!

58. Everything that was the valley, the Awesome Good included is reduced in size and is located down within the top!

59. This completely incredible area of intense solid lasting light grows at it base with every newly arrived area of light.

60. The staring contest and the situation for the viewer grows more complicated as the Awesome Good nears!

61. An impossible indescribable situation that somehow really is happening is now rapidly closing on your position!

62. Real light from the real rock of the Moon makes this growing bending column of rock look very real.

63. Phase two. A rapidly rising bending curvingly strait rock mountain travels toward you right up through space!

64. Down within the rock mountain's upper structure what should be completely impossible is seen!

65. The Awesome Good still stares at you but it has evolved and has it's new second look or appearance.

66. The other main shadows have also sculpted familiar forms. All of these forms are similar and familiar looking.

67. The valley is still there down within the very top of rising column of the lasting and very incredible Moon light.

68. The valley is now seen from a higher angle during the mountain phase when compared to the earlier valley phase.

69. The valley now looks like a sunken room. The flat area first seen around the Awesome Good is the floor of the room.

70. I attempt to describe the sights contained down within the mountain's upper structure in more detail elsewhere.

71. I also attempt to provide more details concerning what the overall scene looks like at this point as well.

72. Describing the light's pressure and intensity is virtually impossible. The intensity and pressure are very strongly felt!

73. Seeing is what it's all about. Once you see, it is very difficult. Keep looking! Stay looking! Don't look away!

74. The actual object is rolling much lower in the scene at this point as it continues moving across to the right.

75. Everything you see that is not the object, but instead is light, is oriented and focused on and moving towards you!

76. The object is moving basically in your direction and the Moon's light's light is definitely moving in your direction

77. At this point my view of this event was about to end. I describe and detail how the staring contest ended for me.

78. Below I expand my return viewer guides in an effort to further describe the various sights that are seen.

79. I hope that all of my efforts taken together will help future viewers see sight of the Awesome Good, and beyond.

A vertically standing crater rim piece positioned on the crossing Object's surface casts a shadow into the Changed Light from the Moon.

The shadow's interactions with the Changed Light cause the light to become shaped into the muscular form of a Man. This shape then lasts as if a statue. This impossibly random series of events is set off in motion by the Ancient Object eclipsing the Moon and it's powerful forces then causing the Moon's newly emerging light to look and behave very differently. Starting with a randomly standing crater rim piece casting a shadow into the great distance of the Moon's Changed Light. The Awesome Good is the Man of Light and the mysterious way the Changed Light imprints itself into everyone who looks. A part of the heart of the ancient mystery is the Changed Light's imprinting effect or Shape Effects, the sight of the incredible fantastic muscular Man of Light standing up and turning around, and the fact that we humans are here on Earth looking up to meet his stare and go face to face with him. A face and the very muscular form of a man that are solid Lasting Light from the Moon, both sculpted, shaped and formed and created by the point very long tall black shadow.

Believing the details I describe is not important.
However having heard the idea of what happens during the ancient Object's return, having read the exact details even without realizing these details are accurate people will be seeing the ancient sights and the ancient sequence of events that I describe and they will be in a situation were they will be realizing that they do indeed have foreknowledge concerning the ancient sights that they will be seeing next as they watch the next Return unfold.

Return Viewer's Guide #2

1. A massive forward rolling crater covered brightly illuminated moon coloured object, rolls down from the area of space above and to the left of the Moon.

2. The object is traveling at a tremendous speed. Once the object enters the area within the distance of the Moon's orbit it will pass down towards the viewer's lower right and disappear below the Earth's horizon as it travels toward the area of space below the Sun in just a very few short minutes at the longest. Exactly how many minutes? I don't know. The object is here within the distance of the Moon's orbit for a very short time as it crosses from the upper left to the lower right.

3. Although I have no idea how accurate this is, I read that in ancient times, using the naked eye, the object was visible for a total of one hour from the time it was first sighted up till the point that it disappeared from view below and to the right of the Earth as it continues down towards the region below the Sun.

4. I believe the object may be slightly taller the way it rolls. The object's surface has an area that I call the raised feature. As the object rolls forward this raised feature comes up from behind the object and travels over the top towards the viewer, then rolls down towards the viewer and goes down under again on it's way around behind and then it comes up from behind the object and then reemerges from over the top of the object once again as the object continues it's constant forward rolling action.

5. The object first covers the upper left corner of the Moon. The lower right area of the Moon is the last part of the Moon to be covered. The object crosses between the Earth and Moon at a very low angle. Because of this low crossing angle, the object has time to roll towards the viewer while continuing to obscure the Moon.

6. During the time that the object is in front of the Moon, before the raised area starts into view over the object's top or northern horizon, I suppose, the object's surface has been "rolling," towards you just the way you would expect a large sphere to do and "look." Now the Moon is completely obscured. Up to this point the object's surface "tilting effect," towards the viewer, has not happened yet. After the raised area has started to into view and then rolls forward beyond or past it's top or

highest position the tilting effect has started or will start very shortly. The way the surface seems to stop rolling and instead looks like a flat area tilting over at you is the point when the valley is seen to the right. The object's surface tilting towards you effect instead of the object's surface rolling at you effect signals the beginning of the valley, phase one of the Changed and Lasting Light event.

7. Even with a small telescope your viewing position is very up close to the massive boulder. My view started between five and maybe seven or eight seconds before the raised area emerged. At this point the Moon was not visible. The object does cross between the Earth and Moon at a low angle and the object does eclipse the Moon for a period of time. The eclipse does last. I am guessing that the Moon somehow could not have been completely covered for more than just a few seconds at this point. One of the factors that I base this on is the way my view started with me finding myself, looking basically looking at the middle, central area of the rolling object.

8. The huge object gracefully rolls towards the viewer, at the same time it was crossing in the telescope's viewfinder, moving, "going," across from left to right and very gently slowly sinking. The object spends quite a few seconds rolling, seemingly, towards the viewer's position while between the Earth and Moon. This eclipse is not some sort of huge rolling object quickly zooming across your view of the Moon from the upper left to the lower right area in the view finder of the telescope, as if it was just simply crossing from the upper left to the lower right of the viewer, and then it's gone. Instead the object spends time rolling at you while it is in front of the Moon. The eclipse lasts.

9. Based on my guess that the full eclipse had probably started only a few seconds before my view started, and based on my guess that only a few seconds passed from the time I first saw the object until the Moon emerges as the valley, I am going to guess that the valley starts approximately eight or ten seconds after the object fully eclipses the Moon. It could very well be a few seconds more.

10. The object continues to basically roll at you. Eventually I did realize that the object was going to miss me once it had moved further to the right in the viewer, eventually passing through the middle of the viewer and after it had dropped down lower. Looking through the telescope, I had no sense of respective, concerning many things including size speed and distance. This is a very difficult thing for me to understand and

explain. I do know that looking through a telescope does effect how one perceives such things as speed and the size or scale of such things as fast moving large object's that are traveling by at a great distance.

11. Before I bent over for my turn, I suddenly noticed that it was as if a bright light had been switched on inside the telescope's view finder. The object's surface itself is so bright that as I was bending over for my next turn, before I actually looked through the telescope, I could clearly see the inside of the eyepiece and how it was constructed and how it looked all screwed together. The inside of the telescope's eyepiece was totally, completely illuminated. After standing, waiting in the dark for my turn the sudden brightness when my view started was a shock for my eye. At the first few seconds at this early point I remembered thinking that my eye was not uncomfortable and I noticed that I could continue to look at the extreme brightness of the object's surface.

12. As the object's raised area comes over the top of the object towards the viewer, the surface suddenly starts to seem as if it is tipping towards you instead of rolling at you. Look right and instantly notice that it is as if someone has magically clicked and dragged the edge of everything that you can see an incredible distance to the right, way way to the right. Phase one of the illusion has begun. It's the Moon but at this point the way the Moon looks has completely changed. Instead of the sight of the Moon, a fantastic incredible valley is seem! It is enclosed by mountains on all sides. The mountains that go across the back of the valley are clearly visible in the extreme distance. Everything is oriented exactly the way you would expect to see some sort of very large valley to look, just the same way you would expect if you were looking at a real valley down here on Earth.

13. As the object continues to roll forward the smooth rounded top of the raised area develops into a curving type of a vertical cliff. Gradually as the top of the raised area continues forward as the object continues to slowly sink, the vertical area right before the top of the raised area, the curving cliff, develops into an overhang. The area under the overhang is were the black spot of the shadow later suddenly develops This takes place in the far left of the overall scene.

14. Down between the mountains of the valley and down over the object's sloping upper right horizon area were you would expect to see some sort of valley floor, you don't see anything at all that looks like a valley floor or ground at this early point. Instead an even level layer of

an effect that looks like the top of a misty white cloud is seen. Everywhere in the valley is like the top of a flat white cloud. The surface or the flat area of what you would expect to see as the valley floor is indiscernible down over the object's far upper right slope or horizon and instead you see down into the area in between the mountains and the smooth level top of a misty white pure clean cloud layer is seen.

15. The light from the surface of the object seems as if it is spread wide open. The light from the raised area that eventually formed the overhang seems to somehow remain in the upper area of the scene and it's as if the raised area has stopped rolling towards you while at the same time the scene in the bottom of the viewfinder expands downward continuing towards you. The effect is like the light from the top of the scene is suddenly remaining in place while at the same time the light from the bottom of the scene continues forward and down. This effect seems to cause the object to suddenly remain in place and at the same time open wide open. A very noticeable valley effect is also now seen to the left in the viewer. The object's surface that precedes the raised area somehow now looks like it is a natural part of the valley as well. The way that the object looks is as if it is some sort of giant snail is extremely pronounced and somehow greatly exaggerated at this point before the "black spot of the shadow," has made it's sudden appearance. Even though the object's surface with it's fantastic craters on the left looks strikingly different than the sight of the immense valley to the right, they seem to flow and blend together naturally and together the left and right become all the same place. At least that's how the effect appeared to me.

16. Suddenly, and it startled me to see something happening, a large black spot appears, clinging to the "ceiling," up under the newly formed overhang. At first I did not realize that the black spot was a shadow. I had never seen a shadow behaving this way before. Because of it's strange unfamiliar action, the way it spread outwards and thickened as it gathered in more ceiling area under the still growing overhang I just watched not knowing what it was.

17. As it turns out the Sun is basically directly behind and I think, slightly above the viewer at this point. The shadow is behaving normally and naturally but I simply had never seen a shadow cast in such an unusual way before. At this point the shadow just does not behave or look the way that you normally are used to seeing a shadow behave. Plus for me at the time with no foreknowledge concerning what I was about to see, I was constantly being taken by surprise by every new sight and

event that occurred.

18. The now very thick and strange looking black shadow very slowly at first, starts to reach and creep a short distance down the cliff face right before it very suddenly drops or even pounces down to the valley floor below. Another unnatural looking action that startled me again! The shadow then immediately starts to race of to the right sweeping across the object's surface. Now suddenly the shadow looked and acted exactly the way that you would expect a shadow to appear and behave. At that point I remember feeling very relieved as I suddenly knew for the first time that I was looking at a shadow.

19. The wide and fast moving shadow looks like a massive black wave as it moves to the right and travels across the object's surface away from the viewer at a constantly increasing steeper angle. Early in the life of the wave of the shadow, before the shadow's leading edge passes off the upper right area of the object's surface and into the light from the Moon, there is time to look down to the ground. Looking left against the movement of the shadow's wave, finding and focusing on one specific spot or area a fantastic effect is noticed. This is the first time that I remember the effect of being drawn closer to the ground , up there. As the shadow reveals new surface details it's as if you are able to see from a closer and closer viewing position. This is the effect that I remember. I can't explain this effect and I did not at first realize this effect was happening when it was happening. I try to go into more detail on the page, "The Shadow." This is the same effect that ultimately draws your eye's viewing position forward to the place were you suddenly find yourself looking down from above and behind at the sight of the back of the Awesome Good right before the Awesome Good turns to it's left and looks up right at you! A truly incredible moment. This is ultimately were the shadow leads your eye and this is ultimately definitely were you want to find yourself seeing from.

20. As the black wave of the shadow races towards the upper right horizon it also seems to start to thicken. As the shadow travels the tops of many crater rim pieces remain in the sun's light above the shadow. Also along the shadow's leading edge many small surface details seem to suddenly reveal themselves. Also shadows are seen crossing and traveling within individual craters. Definitely a feast for the eye.

21. The wave of the shadow does last but overall from the time the shadow actually drops and then crosses over the object's surface turning

the brilliantly shinning object black, until the shadow passes over and then off the object's upper right sloping horizon into the light of the Moon that is there, the valley, only a very short period of time passes. Maybe only ten to fifteen seconds. I always struggle in the area of guessing the amount of time all of the different fast paced separate events take or last for.

22. Certain things are basically simple and easy to describe and many other things are very difficult to describe. Once the shadow(s) pass over the object's far slope into the changed light from the Moon describing what things look like become suddenly more difficult compared to earlier when the shadow was still crossing the surface of the actual object.

23. As the wave of the shadow travels a series of tall individual shadows emerge beyond the shadows edge. These individual shadows travel further and faster than the edge of the shadows wave. As the wave of the shadow passes over the object's upper right horizon these individual are clearly noticeable and are clearly defined.

24. At first the individual shadows are short. I know that scale is very hard to judge and as a result the way these individual shadows looked to me, or rather, what they reminded me of, changed as I glanced to the left away from the series of individual shadow and then to the right, back towards the individual shadows.

25. Once the wave of the shadow has passed over the surface of the object and has reached the valley area that is the changed light from the Moon, the separate individual shadows that precede the wave of the shadow reach further faster into the changed light from the surface of the Moon and it's as if these individual shadows are all now traveling on a flat smooth surface.

26. At first while these individual shadows were relatively short they looked very much like a fence stretched across a piece of property. Somehow the changed light from the Moon's surface appears or arrives simultaneously when the points of these individual shadows arrive. As these individual shadows lengthen they travel over a flat smooth area that is just somehow suddenly there for the shadows to travel over. This flat smooth ground or area forms or builds out away from the viewer farther and farther transforming the wide flat white cloudy misty area that is surrounded by the steep faced mountains of the valley as it travels. A very incredible sight.

27. When I saw the individual shadows for the first time I did not actually realize that they were lengthening or traveling away from me. I was looking all around but mainly I was looking to the right and to the left.
If I had remained focused on the individual shadows I would have seen them growing longer as they traveled away from me. Instead I looked back to the left away from the individual shadows and then I looked back to the right again and saw these individual shadows at a later longer "taller," stage in their growth or evolution. Now instead of looking like a series of fence posts across a yard or property they looks exactly like telephone poles going out across the flat countryside. You see down to this scene from above.

28. In between the time that these individual shadows are short having first emerged right after the big main wave of the shadow has finished passing over and then off the object down over the far upper right slope into or onto the area of the flat white cloud effect and up to and perhaps just slightly past the time that these individual shadows look like a line of telephone poles, there is some time to look at the far upper right slope or horizon of the object. I think of and call this area the far slope.

29. It turns out that looking away from the individual shadows, the "main shadows," goes against the big main point of it all but there is a few seconds, a very few seconds available to the viewer to look over and at the fantastic scene that is there to be seen down over and actually above the far slope. A very important point to remember is that you want to find the point of the biggest tallest widest shadow so that you can let your eye be drawn deeper into the changed light from the Moon. At this point in time that is the big main most important goal. Finding this biggest tallest widest obelisk shaped shadow's point and letting it draw your eye along with it will allow you to arrive at the place were you see down from just above and from just behind, very close to the ground at the exact instant in time that the back of the Awesome Good is first seen suddenly formed having just been created, sculpted by the point of the shadow that was drawing your eye's viewing position forward. Arriving at this point is obviously every viewer's main most important goal! If the point of the big main shadow "strikes," and starts to draw the squiggly line and you are caught of guard looking to the left, the wrong way... let's just say that you don't want that to happen to you...!

30. Everyone should glance at the far slope. It is a sight that is not to be missed but it is not the biggest most important sight that is to be seen plain and simple although this sight is important for many reasons. After

the wave of the shadow passes over the object and the sudden thickness of the shadow is seen over the object's upper right horizon there is time to look at the fantastic brilliantly illuminated shapes that are suddenly seen perfectly arranged on top of and along the line of the top of the shadow. An incredible fantastic scene! I use to think that these incredible shapes were just the tops of some of the standing crater rim pieces and this may be true. Now I am wondering if these brilliantly illuminated shapes could be a combination of the tops of crater rim pieces as well as the light from stars in deep space in the background that is seen over the objects surface. In particular there are two attention grabbing triangle shapes that are the two most prominent and largest shapes that are seen on the top of the thickness of the very black shadow's line.

- Are these incredible shapes a combination of the tops of crater rim pieces as well as background stars? Recently I have heard about the way that planets come into some sort of alignment or grouping.
- Could some of these incredible shapes that are seen above the shadow down over the object's far slope actually be planets?
- What about the new planet that was recently discovered?
- How does that body and or other similar types of perhaps undiscovered bodies figure into this situation?
- What does the big planetary alignment picture really look like?

Time will tell. All of those big important experts out there are really going to love this one! I can't wait to hear what they will have to say.

31. Anyways the point is look, at the shapes above the shadow over the far slope, but don't stay looking at the far slope for to long. If you do stay looking at this amazing scene for more than just a very few seconds it could and will cost you big time. It's like watching the big game. You can be watching and intensely looking and at the same time you can still miss the big play.

32. The most important point is, out of all of the individual tall narrow obelisk shaped shadows there is a single most important tallest biggest widest most important obelisk shaped shadow.

Every viewer needs to locate this shadow at the earliest point possible and every viewer has to focus on the very point of this all important shadow so that they will arrive at that very special place were you suddenly see down from above and behind so that they can see the big sight of the Awesome Good, the muscular Ancient Man of Light.

Then if you are suddenly looking down at the back of the Awesome

Good then split seconds later it will be your turn and you will suddenly find yourself going face to face with the actual Big Thing! the intensely mind bending impossible sight of the Awesome Good!!!

33. Also, if you are looking to the left in the scene and you are looking at the actual object itself be aware of this, the Object will have traveled further down as well it will be further to the lower right. The light from the Moon and perhaps even some of the light from the object remains above the forward rolling object. The raised surface feature that earlier formed the smooth cliff and then the overhang has rolled farther forward and down. The sinking object is literally falling away from the face of the Moon at this point although the object's light remains above the actual horizon of the object, mixed and blended with the arriving light from the Moon. This is the point when the object reaches a certain lower point were it starts to look more like the object that you saw before the view of the valley presented itself. This very unusual visual effect is hard to describe however you will know exactly what I am describing as soon as you see this effect were you suddenly see the wheel turning again. Suddenly you will be able to see the normal round or spherical shape of the ball of the Object again.

34. Every viewer needs to know these next important points. These points that follow are important because every viewer needs to know when the last few seconds are ticking down.

35. These next points signal that time is about to run out. If you see a great expanse of the object's normal spherical curving surface area suddenly visible over, and past and beyond the raised surface area that cast the all important shadow in the first place, time has very nearly actually run out. If you see this indicator, you are looking to the left in the scene, the wrong way at the wrong thing. You are looking at the hard ground of the object. You are looking to the left.

36. Instead of looking left you really really need to look way way back to the right. Look to the right deep into the distance of the changed light and find the point of the big main central shadow because it is nearly time for the first of the ultimate sights. There are very literally only seconds left and you need to know that there is no more time for sight seeing and instead it's time to focus and brace yourself for what is about to happen before your eyes in and then very suddenly literally out of the changed light from the Moon! You don't want to be this close and miss the first instant because you were looking to the left instead of to the

right! Look to the right ... Now!!!

37. THIS IS CRITICALLY IMPORTANT !!!
Again...
If you are looking left at the forward rolling crossing sinking object itself and you find yourself suddenly seeing over and past the object's raised surface area, the overhang, and you are looking at "normal," surface area as described directly above, you have only a very few seconds left to look back to the right so that you will be looking the right way in time to see the sight of the Awesome Good suddenly formed! Look right! Perhaps glance at the far slope for a split second on your way back to the right as a way to ensure that you see the two brilliant triangle shapes in the foreground of this area. It may turn out that these two triangle shapes, the largest most brilliant attention grabbing sights seen above the far slope may not make an appearance or present them selves until a later stage in the evolution of this event. However at this point, you really are taking a big chance if you linger. I would recommend against looking at the far slope for more than a split second at this point. I can't be exactly sure just how many fractions of a second are available for sight seeing in the region of the far slope at this point.

I saw this many years ago and trying to recall the timing of events down to the last split seconds at this critical point is very difficult.

38. Once you can see over and past the top of the overhang, the raised area you do have time to "glance," at the far slope as your eye travels to the right, but that's all the time that's available for viewing the shapes above the shadow down the far slope at this point. I can guarantee these preceding points. There may well be a few more seconds for sight seeing at this point I'm not sure. I can only strongly suggest to you that the time for casually looking around is over. Hurry, hurry, hurry and get busy looking to the right and find and focus on the the big main obelisk shaped shadow's point! Notice that from nearly directly below your eye's viewing position you will see the wide line of the big pointy and straight main obelisk shaped black shadow now very rapidly racing into the distance of the changed light from the Moon.

39. A main most important group of three shadows have emerged and together they travel farther faster than the other individual main shadows. At this point I don't even remember being able to even see the other individual shadows. It really was as if the view of these three main shadows was somehow magnified and the effect was they were the only

shadows that I could see. It's a natural funneling focusing effect that happens as long as you are actually looking. It may be that if you are not looking into the right area within the light, you may not find yourself drawn forward up, to were you see down from. A very difficult effect to describe in spite of the fact that this effect is so vividly and easily remembered. Very soon everyone who also sees, will know exactly what I mean. If you are looking to the right and you are using a telescope you should be able to easily find the big main shadow because as I describe at this point everything just seems to start to happen naturally. I think that if you are looking to the right it would actually be hard to miss the sights I describe. You will not find yourself doing any sort of "searching," for the big main shadow at this point instead if you are looking right, the light will allow you to see because of the focusing funneling effect that I describe.

40. Now at this point the position that your eye is seeing from is lowering down closer to the ground along with seemingly following along behind the three main shadows as if being drawn forward by or even riding forward in the light. Also everything that is out wider that is a part of your peripheral vision seems to naturally follow the flow forward. Now you will have noticed that during the last few seconds, the flat ground is closer below you as your eye's viewing position seems to race seemingly faster, farther forward as it keeps gradually getting closer to the ground.

41. I really don't know if the point of the big main central shadow is actually accelerating at this point or not. It is definitely moving away very rapidly but is it also accelerating? I just don't know. I think of it as accelerating but I don't actually know if the point of the big main black shadow does accelerate or if it is just simply moving very fast at this late stage.

42. By now everyone needs to have found the big main shadow so that they will be able to see the first instant in the Moons changed light. Do not look away! You would think that not looking away would be simple. Brace yourself for the very complicated difficult sights that you will be seeing very shortly! At this point there is nothing more important that focusing on the point of the big main shadow so that, you are ready to see down from above and behind. This really is the place were you want and even need to be seeing from! There are now fewer split seconds left... Time now runs out!

43. This is it!
Suddenly the central taller longer wider shadow's point goes up and down and up and down, then instantly the shadow's point cuts zig zagging to the left and to the right and it starts to draw a squiggly line that is the shaping and and the forming and the sculpting and the creating that produces the sight of the back of the Awesome Good!!!

44. This really is it! The first great sight! Your eye has arrived at the place were you see down from! Look down and see the clear awesome sight of the very muscular back of the Awesome Good! Fantastic!

45. This moment only lasts a very few short seconds before it passes and it's over. An incredibly peaceful and tranquil moment of thought incredible!

46. Our ancient ancestors knew this sight and I can only suppose that this is why we all also know this sight and this pose. This is obviously were this ancient depiction comes from. Enjoy this very very brief sight because this is all about to end in the most stunning shocking manner!

45. Shortly after the next few seconds you will be suddenly even physically forced to realized what you are suddenly seeing. This is the point that is most difficult to actually realize because of the impossibly shocking nature of the next sight. At the same time knowing ahead of time that you are not suppose to feel extreme shock and fear may very well help you at this point when you really do see the heart of the changed light for the first time!

46. This is only my opinion, but if a person is able to remember that somehow it's all about the opposite of shock and fear then perhaps they may be able to experience less shock and fear during their own view of this event when they are actually really seeing in the ancient way and also importantly later after their view of this event is over when they are remembering this sight the special way that the shape effects work throughout all of this. It may turn out that the more you are able to feel and think in terms of overwhelming joy and love during those very personal face to face seconds of the Big Noticing the more likely it may be that those are the feelings that you will instantly feel whenever you are actually really thinking about this experience and this face to face sight later when very suddenly you see and feel that the shape effects are at work.

47. I do and at the same time, I don't actually know this because instead for me, it's the extreme opposite situation. I had no idea that it's all about extreme joy and love. Instead I had the extreme opposite reaction when I saw and went face to face with the big sight and as a result, unfortunately for me when I remember the special way that the shape effects allow you to remember, I suddenly found myself very involuntarily feeling the exact opposite way that you are suppose to feel. A very difficult effect to describe but this effect is real. This is very obvious to me. I am hoping that there may be a chance that mentioning this very difficult and personal aspect of this experience may help others to know and experience the fantastic face to face seconds the way I truly believe they were intended to be, by nature. For me being this open and honest is difficult. As a part of trying to do and say what is right I clearly understand that I have to let the things I know lead me even when some of those things lead directly into personal areas that are difficult to openly talk about. If it turns out that including the above points and comments actually helps people to know this experience the way I think it may be intended to be experienced, then I am glad. The points I mention above are the single most important points that I go over with my two young children. I want both of them to see and I am hoping that they will only know and feel the joy and love.

48. The peaceful tranquil few seconds that go by while you look down from above and behind at the back of the Awesome Good very suddenly end! Basically you have only had enough time while looking down from above and behind to realize what it is you are looking at down there before suddenly everything that you are seeing drastically changes. I am able to read about many other big moments that follow this particular point in time but except for a few, the rest of the big moments or big sights, all happen later during phase three of the changed light illusion event, after the changed light arrives at or apparently very near the Earth. The face to face moment of the "Big Noticing," is easily the biggest most fantastic and most difficult and wonderful sight that I have ever seen.

49. My viewers guide will have hopefully helped you to arrive at this incredible fantastic place in time and space. Now it's very much up to you to keep looking so that you can go as far past this point as possible. I was very fortunate to arrive here so that I could experience the actual face to face moment. If you are finding yourself looking at the back of the Awesome Good then you are also going to suddenly be noticed! A very fantastic moment when you see the most spectacular sight. I hope that the big main point that I make from this special moment onwards

will be of some help. Simply put, keep looking! Don't look away! Focus on forcing yourself to return the intensity of the stare! Remember the part about joy and love!

50. Suddenly in only a split second, the Awesome Good turns to it's left, spins to it's left and gets up all in one motion and is suddenly oriented directly at you and the ultimate face to face staring contest is on!!!

51. A number of heart stopping seconds go by as the face to face seconds last! During this time the next one way arriving wave of light moves up in your direction from the Moon and this causes the face to face to drift up closer to were you are looking down from. The distance closes as the Awesome Good rises and nears you! Definitely incredible wonderful and fantastic come nowhere close to describing it! Suddenly these face to face seconds end.

52. Now the very bizarre transition period after the valley phase and before the mountain phase occurs. The outer edges of the light from the valley pulse inwards a number of times. The Awesome Good itself wiggles around and orients itself differently and is evolving at this point into it's second evolution. A very bizarre sight that I can describe in detail. Keep staring back, and don't let the continuing face to face staring contest that you are locked into at this point end!

53. The place were you see from has raced back during the light's inward pulsing then almost instantly the intense pressure and force of the lasting light's focus arrives and is felt right after the light's inward pulsing has stopped. The difficulty for the viewer constantly keeps going straight up!

54. Everything is about to drastically speed up and then incredibly everything that you see will be rapidly getting closer to you as once again incredibly the intensity and the pressure of the light's focusing stare starts to build higher and literally rapidly rise right up through space as the actual Big Thing the Man of Light seems to be noticing and staring at you even more intensely. Stay looking although now changed the incredible face to face sight of the Awesome Good is still happening and is rapidly nearing and closing in on your exact position from down inside the incredible growing bending curving mountain of changed Moon light, as every viewer will very soon suddenly see and realize.
Stay looking! Don't look away! The longer you can see for the better you will feel about it later. My sincerest best wishes to you and your family from me and my family! Good Luck!

Return Viewer's Guide #3

1. When I try to describe the object's motion and direction it is always from my own viewing position and perspective. The place were I stood in Mississauga Ontario Canada. The forward rolling crater covered object arrives from the area of space above and to the left of the Moon. The first part of the Moon that is covered by the object, is the Moon's upper left corner. The object appears to roll down from that area of space. Overall the object is moving at a tremendous speed and is between the Earth and Moon for only a short time. Because the object's crossing angle is very low, the object is between the Earth and Moon, rolling towards the Earth, for a fairly long period of time, relative to it's very high speed.

2. Although huge and massive, the object is smaller than the Moon. I understand that something that is smaller than the Moon can still in fact be very large and massive. The object is easily in fact a very very big place. The object crosses the line of sight at a position between the Earth and Moon, well out in front of the Moon. This crossing point and angle allows the object to completely obscure the larger Moon for a many seconds as the object continues to roll towards you. A normal in focus view of the Moon is much less clear than the brilliantly shinning, incredibly clear, extremely in focus view of the object and the object's surface details.

3. The object's crossing angle is so low that it certainly does appear to "the viewer," that the object is rolling basically in there direction. The object does cross at a point that must be a great distance, well out from the Earth as well as the Moon. Through a telescope the object appears to be very close and bearing down on the viewer as it seems as though it surly must just be somewhere close and directly overhead. The telescope places your eye's viewing perspective at a point that is well out from the Earth. Your sense of being at a safe distance back from the forward rolling object is completely gone as the extremely bright massive rolling place fills the telescope's eye piece.

4. The object is slowly sinking as it moves from the left to the right in the viewer. The object's left to right crossing rate of speed is higher compared to it's much slower downward sinking rate of speed. An incredible hair raising sight that is guaranteed to get all of your adrenaline pumping!

Even though the object does miss the obvious startling feeling that you get from viewing through a telescope is that basically it's headed for you. Looking through a telescope if you did not know ahead of time that the object "does in fact, miss the Earth," you would think that surly, it was definitely going to hit. For me at the time not only did I think that it was going to hit the Earth, at first I figured it was actually going to hit my neighborhood! The first point that I had any sense at all that the big rolling place was going to miss was when it finally started to fall away from the Moon down to my lower right once it had passed the center of the view through the telescope. At that time however, as a kid, I only knew that it had actually missed the Earth completely, once days had pasted and nothing seemed to have happened. I know that this has got to sound completely ridiculous, but I was a kid looking through a telescope that night for the first time and I had basically less than zero overall understanding. The view through the telescope caused any and all sense of perspective to be completely thrown off. I had no idea or sense that the object was actually far away and crossing at a point well out from the earth. I am easily able to understand that this basic point is an important point to know, and may actually help future viewer's during these very stressful moments that are approaching us right now.

5. All future viewers will know that their telescopes are still pointed at the Moon. That is something that I did not know when I bent down and saw the object for the first time. Instantly I thought the telescope had been moved. While I was seeing I had no idea that the Moon was about to emerge from over the object's right area and upper right horizon. - I try to take time as I write to explain how various factors like the one I just mentioned helped to cause my, and contributed to my lack of perspective and my very very limited understanding concerning what I was seeing, when I was seeing it. I what to try to ensure that future viewer's are in a better position than I was concerning understanding what they are seeing when they are seeing it. I can only do so much but I am going to keep trying my best to help future viewers. Looking through a telescope throws off your sense of perspective concerning the size, and distance and the speed of the object and the Moon's light. That is unavoidable. What is avoidable is future viewer's having a complete lack of perspective and understanding when they are watching the return events unfold. A lot happens in a very short time. The light that you know from the Moon becomes changed and unrecognizable and it will confuse and fool you. I do understand that I am able to help people understand some very basic things now that I understand these basic things better.

6. Except for the left side and perhaps the right side of the forward rolling object, (I did not see or rather I could not see the actual right side of the object. I try to explain.) the surface of the object is completely covered with even hundreds of the most fantastic and unique craters. Individual vertical standing tall narrow crater rim pieces very neatly circle all of the various size craters. The standing crater rim pieces have a very uniform look to them. Their thickness may reflect the thickness of the object's crust.

7. The craters themselves are not some sort of deep hole the way you would normally picture a crater like for instance a typical lunar crater. Instead all the craters that I noticed and took a look down to were no deeper than the thickness of the crater rim pieces. It may be possible that the area inside the circle of the craters may be basically flush and level with the object's uncratered surface area. I am unable to be more precise. My guess is that the craters are as deep as the crust's thickness the thickness of the standing crater rim pieces. I know that it is one or the other.

I remember this detail and I see this detail being a certain way. I do realize that I am pushing my memory to it's limit in this case and I really don't know which situation is the case. The sight of the crater's are real and not a shape or form made out of changed light. As a result I have to try to simply remember the details concerning the depth of the craters as I describe basically without the benefit of the way the shape effects lets you remember details. Aside from this area concerning the depth of the craters were I have it narrowed down, the rest of the descriptions I give concerning the craters seen on the object's surface are completely accurate, guaranteed. If you've ever seen a standing stone circle down here on the Earth and we all have then basically you've also seen a crater on the returning object, and that's also guaranteed!

8. Once the sight of the Moon is obscured by the object all you see is the massive crater covered boulder very gracefully rolling at you. Suddenly a large raised area comes into view from over the top of the forward rolling object. Future viewers will have already seen the raised area coming over the top of the object towards then as they will be looking at the object before it crosses between the Earth and Moon and then blocks out or eclipses the Moon. Does a shadow develop with each of these earlier rolls as the object approaches it's position between the Earth and Moon? I don't know because my view of the returning object started after the raised area had gone down and around and was getting ready to once

again come up from over the top again once the object was completely obscuring the Moon. It could be possible that from whatever point onward once the object has come down low enough, a shadow may be cast from under the overhang that the raised area becomes with every forward rolling revolution that the object makes before the object arrives in place in front of the Moon or maybe not I don't know, we shall see.

9. My view started seconds before the object seemed to magically settle into place in front of the Moon. Now I understand that this settling in starts when the light blending between the object and the Moon starts to happen. At first I saw the object rolling down without any sort of place attached to the right. A very fast paced series of events take place. Once the object is in front of the Moon, each part of what you are seeing is an area or time segment or phase, a sight, that lasts for only a few seconds.
Some things take longer to happen that others but the things that happen, the separate sights only last for short seconds. I try to estimate and remember how long the various sights lasted. It has to be remembered that I can't be completely accurate down to the last second. That sort of accuracy guessing time and counting seconds from so long ago is something that is impossible to do with complete accuracy. I also understand that the shocking nature of the experience after, and while the Return events are happening, makes judging time down to exact seconds very difficult. I have to say that I think that I am basically fairly accurate overall. Being accurate overall does not mean 100% accuracy. There are sights that last for split seconds, they last for split seconds. Things, sights that last for a much longer time are much harder to accurately estimate.

10. Suddenly the large raised area comes into view from over the top of the forward rolling slowly sinking crossing object. The object's entire surface area still seems to be rolling toward you just the way that you would expect. Then suddenly the area before the high raised area seems to suddenly be flattish looking and tilting toward you as you watch instead of the normal rolling toward you that you had been watching up to this point. The sudden tilting is the point were the Moon is suddenly seen to the right and to the upper right past the edge of the object's rolling sinking horizon. The light from the object's right and especially the edge of the object's upper right horizon, is suddenly blended together with the now changed light from the Moon. The incredible sight of an immense valley is seen to the right. It's the Moon but it does not look at all like the Moon anymore. Instead a fantastically clear view of a very real looking valley is seen.

11. Steep mountains are seen on both sides with a wall of mountains going across in the distance. Before the wave of the big shadow that is about to be cast under the raised area's growing overhang forms, past the object's far upper right downward sloping horizon, the area of the valley is as if a flat white cloud. A white cloudy misty look is seen were you would expect to see the valley floor. This white misty cloudy look is seen valley wide and covers the entire valley floor area from the edge of the object's upper right downward sloping horizon right up to the edges of the mountain's smallish foothill type of look that is seen at the base of the steep face mountains.

12. As the object continues to roll forward and slowly sink and move from the upper left to the lower right in the telescope's viewer, the very smooth and curving raised area begins to gradually form a smooth curving vertical, rounded at the top, rock cliff. Then an overhang gradually forms as the object's raised surface feature continues to travel over forward and then down towards the viewer.

13. The Sun, and the Earth, and the object, and the Moon are all lining up. As the big raised area continues forward it continues to form and develop into more of an overhang. Very suddenly a very noticeable, very black shadow appears up under the overhang. A very big and very noticeable black spot. I think of it and refer to it as the black spot of the shadow.

14. The black spot that I didn't realize was a shadow when it first appeared, instantly starts to progressively gather in more of the ceiling area under the overhang to match the object's forward moving, forward rolling sinking motion. During this time the shadow thickens and/or gets deeper I suppose depending on how you look at it or depending on how it happens to look to you. A very bizarre unnaturally behaving shadow. I suppose the shadow seemed unnatural because I had simply never seen a shadow formed in this way before.

15. The black spot of the shadow's sudden appearance instantly caught my attention when it appeared in the left area in the overall scene. The black spot of the shadow spreads and thickens. Suddenly the shadow seems as if it leaps downward to the valley floor below. At this point even though it's as if the shadow has dropped down to the left end of the valley, the shadow is actually still being case on the actual surface of the object itself. The shadow now an awesome wave still has to cross over the object's surface before it actually reaches the object's upper right

horizon, the area I think of and call the "far slope." This takes some time. The shadow does travel fast but overall the shadow does has a considerable distance to travel.

16. Once the shadow reaches over and off the object past the far slope into/onto the area of the flat white clouds somehow it's as if the light from the Moon were the flat white clouds are turns into smooth flat hard rock, to meet the leading parts of the shadow's edge. A very natural transition between the actual hard surface of the object and the light from the Moon is seen. It's possible that there may be a slight delay between the shadow traveling across the object and then the shadow traveling along the flat rock or ground looking effect that the flat clouds turns into. I was looking left and right and I did not follow and look at the actual edge of the shadow's wave in an uninterrupted manner. Some things are unclear to me because of the way I was looking back and forth to the left and right.

17. Once the wave of the shadow has passed over the far slope into the valley area that is surrounded by the mountains there is a fantastic scene above the thickness of the line of the shadow. Many many shapes are brilliantly illuminated above the thickness of the black shadow. Just to start will, the tops of crater rim pieces that remained above the depth of the shadow account for the majority of the fantastic brilliant shapes that are seen down the far slope. In a very bizarre way these shapes seen as if they are all situated so that they are also orientated facing and looking out into the distance of the valley just like you are. Another strange effect is that even though obviously the object itself is still rolling and sinking the shapes that are seen down the far slope remain unmoving in place. As a part of the light blending and the effects of the changed light somehow the light from the tops of crater rim pieces that were there is still there. Obviously the craters are continuing down and are literally down below and gone but the light that reflected off the tops of the rim pieces is is still there and now a natural part of the scene.

18. Also it could be possible that the light from back ground stars and perhaps even planets that may be in some sort of grouping could also contribute to the what the viewer sees. Maybe once the surface of the object has dropped down to a lower point light from more stars may suddenly emerge and become part of the fantastic array of brilliant shapes seen down over the far slope. In particular there are two fantastic triangle shapes that are seen within this area. I am uncertain in regards to the specific timing of their emergence.

- Were these two triangle shapes visible or there and developing or evolving at the same point that all of the rest of these brilliant shapes were first seen above the shadow?
- Did these two triangle shapes appear at a later point after the first group of shapes are visible above the shadow?
- I don't know because I was looking left and right and even up but especially down to the ground, up there. My feeling is that they appeared later after the first group of shapes above the shadow. I also think that because of the way they really were extra stunning and larger than the other shapes and because of their possible later appearances they may well be a pair of stars instead of reflected light from the tops of crater rim pieces. I don't know this, instead this is a feeling, a guess. Later we will see how all of this actually works..

19. Overall once the shadow has passed over and off the object into the light of the Moon, the valley, there is very little time left to do anything other than focusing on finding the big main shadows point. Finding and following the big main shadows point is the most important task for the viewer at this point. The seconds are very quickly counting down and the big point in time when you are going to be looking down from above and behind at the very end of the black wide long road that you have been hopefully traveling on is approaching very fast.
Time is running out!
Look to the right and find the main group of three shadows that are traveling together. The middle shadow is wider and longer or taller than the two outer shadows that trail behind the main central shadow. Find and focus on the very point of the big main shadow!It does travel on a slight angle even though it is actually straight.

20. The object is crossing from the left to the right out in front of you between the Earth and Moon. The Sun, the actual real Sun is positioned behind the viewer behind the Earth in line with the Moon and the object and the Earth. At first the shadow travels from left to right and at the same time the shadow is traveling on an angle away from you that is steadily increasing as the shadows lengthen. Initially there is a lot of left to right motion. Later as the lines of the shadow grow very long the angle they are seen at has increased greatly and now the shadow is lengthening and traveling almost all the way up to ninety degrees or straight away from you. As the object continues to cross the shadow goes from having a lot of left to right motion all the way to nearly traveling straight away from you without any motion from the left to the right at all. The lines of the three main shadows become straight as they pass over the flat

whiteness of the valley turning the the flat white cloudy mistiness into flat smooth hard ground.

21. Does the shadow reach all the way to ninety degrees before it starts to go up and down and up and down? I don't know. It might, it might not. I don't really think that it does but I can't be 100% certain. Events start to happen very fast at this point. This is the point were the viewer at least this viewer started to lose track of the details of exactly what's happening. My viewing position was racing forward as if zooming like a zoom lens up into the distance. I was less aware of the surrounding details all the time as the point of the big main shadow dragged the place I was seeing from further forward into the distance of the Moon's changed light. In general I hope that my descriptions can help future viewer's find their way to this point. The exact angle of the shadow is really not important at this point. The point is to find and stare at the very end of the tallest widest shadow that is in the right side of the entire scene that you are able to see. That's what's important!

22. I spend time mentioning smaller details and points because I understand that mentioning these points may turn out to be useful for future viewers. Not because the small details are important or critical to being able to find the big main shadow. If you are looking the correct way to the right at the right time, once the shadow has traveled over and off the far slope you will see the big main shadow!
It will just simply be there right in front of you for you to follow.

23. If it turns out that the line of the big main shadow goes past ninety degrees before it strikes and it might who cares? This is a general viewer's guide. I will be out or not 100% completely accurate some of the time. Being completely 100% accurate concerning every detail all the time is not possible. In general following my viewing tips and suggestions will get you to the ultimate ancient place in the Moon's changed light when it's time to look down from above and behind and it's nearly time for the Awesome Good itself, to turn around and get up and notice you. This is were you want to be! Me being out a few degrees here and a few degrees there will make zero difference. Follow the points that I suggest and you will get noticed!

24. I mention the idea about the timing of events. I mention about seeing over and past the object's raised area and how it is important to know that if you can actually see over the object's raised area and you can see the normal round spherical shape of the object again there are only seconds

left before the point of the big main shadow starts to sculpt the form of the Awesome Good in the right area or the upper right area of the valley. This is actually important to know but if you are already following the wave of the shadow into the distance to the right and you are not looking back to the left and you don't plan to look back to the left then suddenly seeing over the top of the object's raised area is not important to you. You will be already looking to the right and by this point you will have easily already found the big main shadow and so whatever is going on way back to the left will not matter to you. You will already be looking at the single most important sight that there is to see at this point.

25. To the left the shadow appears under the newly formed overhang. Then suddenly the shadow drops to the surface of the object below. Then instantly the shadow races from the left to the right over and off the object's surface towards and into the upper right area. Tall individual shadows separate themselves out ahead of the big black wave of the shadow. Find and focus on the point of the big main tallest shadow and don't look away as it grows longer and longer and then suddenly your eye will arrive at the place were you see down from and then you will see the muscular back of the Awesome Good. Then you will see out to the edge of the flatness of the ancient place in the Moon's changed light that surrounds the place were the Awesome Good is positioned. Then the Awesome Good turns around and you will get noticed! It's basically very simple. Follow the shadow from the left to the right and you will see! Then beyond this point it's all about not looking away! Stay looking and you will see more and as a result, you will know more! Then everything speeds up, and the intense pressure arrives and builds and then everything goes from there! Be happy be glad, and try to remember the part about joy and love. In the end joy and love matter! Once the face to face situation is in place it keeps happening! The inward pulsing starts and ends and then everything that is light just simply rises!

Don't look away! Keep forcing yourself to receive the light's indescribable pressure and very difficult intensity! There are a lot of details that are worth mentioning but overall it's about looking at the right place at the right time. It's a natural!

Return Viewer's Guide #4

1. We were using a small backyard type of telescope. The telescope was new and of a decent but I suppose average quality.

2. We were using whatever eye piece gave us the overall clearest view of the Moon. It is my opinion that this overall clear view of the Moon contributed to the crystal clear very in focus view of the object and it's surface, and the extremely clear sight of the Moon's changed light.

3. Basically I think that you don't want to be in really close to the Moon. I believe an overview type of viewing perspective will prove to be best overall as opposed to a very tight in close view of the Moon. Once again this is only my opinion. If a person had the luxury of more than one telescope then perhaps both the in close and an overview type of viewing perspective could be used to great advantage. I think that this would only be the case if the viewer understood the fast paced timing of the sequence of events.

4. The overview type of viewing perspective is the type of view I had. This allowed me to see the overall scene. Once the object had settled into place in front of the Moon, I was able to see space to the left of the object and I was also able to see over the top of the object.

5. With the Moon positioned slightly to the right in the viewer everything seemed to be framed perfectly. The object on the left and the view of the Moon, the background for the valley on the right. If the Moon had been perfectly centered in the viewer then there could be a chance that the object's left horizon, the left side of the object, along with some of the object's surface, may not have been visible to me.

6. The higher power small backyard telescope that provides a much closer look at the Moon may not give a viewer a chance to see the overall scene quite as well as the lower power telescope. On the other hand a closer tighter view of the Moon itself would allow for a very in close view of the spot were the Man of Light is created and formed by the point of the big main shadow.

7. For me personally I understand that the overview type of view provides an extremely in close view of this incredible event and the incredible sights that form there. At that very special point in time your viewing perspective seems to have raced forward and you seem to be

looking down at the sight of the Awesome Good from an extremely close viewing position. Defiantly for me at least, this is close enough! Seeing from even closer is something I think of as being the realm of only the very brave!

8. The idea of switching from a lower power telescope to a higher power telescope once the time is right is something I have thought of but I seriously doubt that I will do.

9. The series of events moves at such a rapid pace that if you are not sure the timing you could get it wrong and you could actually miss the ultimate sight at that ultimate instant in time. Defiantly viewers don't want this to happen to them.

10. My main advice is once you are watching the shadow's awesome wave and you have located the big main obelisk shaped shadow you should remain focused on the idea of concentrating on the point of the big main shadow. The idea of stopping to look through a different telescope, could turn out to be your ultimate life's regret. I can only recommend against it. Once you are actually seeing the object's awesome surface with the shadow's wave moving across it keep looking. You are almost there and very shortly you will be seeing the awesome visual spectacle that follows. Don't look away! Locating the big main shadow in the middle of the three main shadows is a natural once you are looking to the right in the right region at the right time.

11. In my opinion, in the end or is it actually the beginning when the Moon's light very suddenly focuses on you remaining looking will be one of the most difficult things that you will ever attempt to do. I do think that if a viewer is prepared and actually knows ahead of time that they will need to "brace" for the big moment, and that there is a big moment, then remaining looking is something that they will be able to do, I hope. I understand that switching telescopes should only be done early in the life of the shadow and not later. When future viewer's are watching the event unfold I hope they will have some sort of sense of were they are within the sequence of events based on my descriptions. Before the shadow it's safe to switch telescopes. Early in the life of the shadow it's also safe to switch telescopes. The shadow has a very short life. For most people there may only be one chance to get this right. Late in the life of the shadow switching telescopes for some sort of closer look could cause you to miss the sight of the big moment. Stay looking and don't let that happen to you.

12. During the last return I had no clue that I was about to see the ultimate sight. I was a young boy with zero experience no lead up and no preparation and completely unsuspecting. I looked down through my friend's new telescope again after I had looked at least once before as I know I had seen the Moon by this time at least once. As if someone had changed the channel without me knowing the massive object was just simply suddenly there. At that point I had no clue that the telescope was actually still pointed at the Moon as I instantly presumed it had been moved and pointed at something else as if my friends Dad had pointed the telescope at the next big thing we were going to look at. Right from the start any sort of perspective or understanding was totally completely thrown off. I was already instantly struggling badly with my first peak at the very intimidating massive crater covered boulder the object that seemed to be basically rolling right at me!

13. That was just the start. I have no idea why I was not already running even at that very early point in time. Like me, surely most or all telescope aided viewers will defiantly have that very uncomfortable caught in the middle of the road feeling! I think that because I knew so little and because I did not know that such a sight was actually unusual somehow I stayed looking from this initial early point right through the series of visual events that takes place all the way through the big instant in time. I also managed to stay looking for a time past the big instant but very unfortunately my view would be soon rapidly nearing an end as I describe. Completely shocked and totally confused and overloaded and badly struggling comes no were close to describing it.

14. Ultimate seconds that everyone should know! It turns out that people can know these seconds and I understand that they will know these seconds. The impossible the indescribable sight will be there happening again soon! People need to understand that they can see it to and if they see it then they will know.

15. In the end it's all about seeing that's the point! The only way that's going to happen is to look and to stay looking. I understand that this can only sound impossible but the simple fact is that unbelievably actually remaining looking is not easy. I try to describe the shocking pressure and strange intensity that is really felt by the viewer. The full force of this pressure and intensity arrives with the second phase of the lasting light event, the rising bending incredible lasting light mountain that happens right after the the inward pulsing of the transition period between the first and the second phase of the Changed Light event.

16. As far as the telescope aided viewers are concerned, once you are watching the return and it's nearly time, as I describe very soon my advice will be only a small help at best. My viewer's guide will get you there but from that special instant onward it's every viewer for themselves. Alone and going face to face with the big sight the actual Big Thing is not a simple thing. It is my hope that along with the knowledge that a special sight is going to be seen created in and out of the distance of the light of the Moon, the idea of remaining focused on not looking away will at least be of some help to all of those who are interested and all of those who are fortunate enough to see and know the very sudden extreme degree of pressure and intensity and difficulty that is faced and actually even felt.

17. Once the object is sighted, know that a very special sight is there to be seen and as always once again somehow it's way way more than that! The ultimate staring contest is way more than just on as now all viewers are themselves realizing that they really are looking at and incredibly seemingly being noticed by the all time ancient mystery. The ancient Man of Light. Force yourself to absorb the pressure that the light brings.
It is good for you although while you are actually feeling the light you may find yourself wondering if it's doing damage internally. The feeling of the movement caused by the light is suppose to happen as far as I understand it. It's not painful and it's not harmful. It isn't even an unpleasant feeling but certainly it is an unusual and very different sensation to feel the changed light find it's way into the place were you are already wired to receive and store it. Hang in there and stay looking!

18. Along with all of the intensity and difficulties that this brings and that I know, certain things turn out to be easy. Like the fact that this happens. For me this is easy. By a total unexplainable completely lucky fluke I saw it. Somehow I saw the big place roll by last time and I saw what the big rolling place does. I can't explain it but I can try to describe the sights that I saw.

I was going to add onto the above viewer's guide and finish it but instead I decided to include it and leave it unfinished just the way it is now. I can only hope that after reading the various descriptions that I am offering one day you will also be able to look back and remember your journey to the end of the Black Road.

I hope that you will see the shadow's point cut left and right zig zagging and drawing the squiggly line that it does draw and actually will draw for you again when you focus and remember. If you are lucky enough to get noticed then you will be able to understand the shape effects especially if you really try hard later. Focus on the Moon's changed light when you are actually seeing the Moon's changed light. Try to realize that when you are suddenly feeling the pressure and indescribable intensity that this is the place that you want to be and it won't last. Push away the fear and the worry and try to think about enjoying the feeling and the sights that you will be seeing.

If I'm lucky enough to be back seeing again these are the types of thoughts that I am going to try to force myself to have. I hope that this is what will happen for me, my two children, and for you and yours!
It is the ancient mystery and it is the sight of the Awesome Good!
Sincerely

Return Viewer's Guide #5

1. Shortly, a very massive crater covered forward rolling moon coloured object is going to roll down across between the Earth and the Moon. The forward rolling, slowly sinking object crosses between the Earth and the Moon moving from the upper left to the lower right.

2. The object is smaller than the Moon. The object crosses at a point between the Moon and Earth at a point far enough out from the Moon towards the Earth so that the object's smaller size, relative to the size of the Moon, is able to completely obscure or eclipse the Moon.

3. At a certain point the total eclipse ends. The Moon's light emerges above the object's upper right horizon and is spread out wide open to the right and upper right of the object. I think of it as "Moon Rise." Obviously the Moon does not actually "rise," instead the crossing object's surface is sinking down allowing the Moon to be seen above it's forward rolling surface and especially to the right were the Moon's light opens up and spreads out.

4. When the object is crossing between the Earth and the Moon the Sun's light that has reflected off the Moon's surface and is now traveling toward the viewer on Earth becomes transformed as it passes over the crossing object's fast moving, forward rolling, slowly sinking surface.
The Moon's light suddenly does not look or behave at all like the sight of the Moon. Some sort of very fantastic effect is generated and/or caused by the crossing object's forces.

5. The object's largest surface feature, as far as I know, is a rounded smoothly curved raised area. It rolls up into view over the top of the forward rolling forward moving slowly sinking object.

6. Suddenly instead of rolling at the viewer, the surface area of the object that precedes the top of the object's raised surface feature seems to start to tip or tilt towards the viewer. This signals the beginning of phase one of the Lasting Light event, The Valley.

7. The fantastic light blending and light changing has started. The light from the top of the object seems to stop and remain while the object itself is obviously still moving and rolling forward. The bottom area or regions of the object continue forward and down at the same time. The snail shape of the object becomes greatly exaggerated. The light from the top

of the object somehow seems to stay where the actual real surface was while the bottom of the object spreads down and the entire scene seems as if it's closer to you as the object itself seems as if it's opening wide. The surface area that precedes the raised area seems as if it expands or lengthens. A very bizarre effect that is very hard to realize when you are actually seeing this effect. Later after the shadow has been cast and the shadow has traveled across the object seconds before the shadow strikes and draws the squiggly line, suddenly the object reaches some sort of lower critical point. At this point the light from the object's upper regions is no longer being left behind and blended and or mixed and or tied into the light in the same way. At this new lower point it is as if the object has suddenly reemerged or fallen away from the Moon and is suddenly seen down lower actually turning or rolling again very similar to or possibly exactly like the way it was rolling before it started to tilt instead of roll at the point were the light blending and the first sight of the valley happened in the first place.

8. I am jumping ahead a little bit here but this is a critical point that needs to be understood concerning the timing of events that lead directly to the place in the light when the point of the big main shadow actually sculpts the form and the shape of the Awesome Good. When the spherical normal curving round surface area is suddenly visible over and past the object's large raised surface feature there are literally only seconds left before the shadow strikes! You are able to suddenly see over and past the raised area once the object itself seems to be fallen away down below in the overall scene as I try to describe. Hurry and look back to the right. If you look back to the right it will be easy for you to see the main group of obelisk shaped shadows racing away from you. Find and focus on the point of the big main middle shadow. This big main shadow is easy to find and see. The central shadow is substantially taller and twice as wide as the two outer shadows that it travels with into the distance.

9. Going back, after the full eclipse, the Moon's light is first seen again over the object's upper right horizon. The Moon's light is fantastically changed. To the right, the view is of a fantastic valley. It's as if the Moon's light is tipped away and spread wide and deep. You see an expansive valley with no real discernible valley floor at this early point. Instead of any sort of actual valley floor at this early point before the wave of the shadow has crossed over the object a flat white cloud looking effect is seen every where in the area of the valley between the mountains. The entire scene is surrounded around it's perimeter with the

Moon's seemingly piled up surface light forming the sight of many very uniform looking steep faced mountains. A normal looking view off into the distance of a valley in a land area is seen. Even the flat while clouds look natural. The view of the valley is oriented exactly the way you would expect a view of a valley to look. A very convincing valley look is seen as if some sort of new place. This is the sight that you will see to the right of the object once the Moon has seemingly risen. It is the Moon but this sight does not actually look at all like the Moon.

10. The rounded raised feature continues rolling forward and slowly sinking. From the viewer's perspective, the object's raised rounded main surface feature starts to develop into a rounded steep cliff. Gradually the smooth vertical cliff becomes an overhang.

11. Suddenly a very noticeable black spot appears under the newly formed overhang. The black spot of the shadow. The Sun is located above and behind the viewer. It begins to cast a shadow under the newly developed overhang. The black spot of the shadow grows. The shadow spreads out and thickens while it seems to remain clinging to the underside of the forward rolling overhang.

12. The shadow is about to drop to down to the ground below, the surface of the object. At first the shadow seems to slowly stretch or reach, gradually spreading down the area under the overhang as it thickens even more..

13. Suddenly the shadow then seems to leap or pounce down to the ground below. The shadow then immediately very rapidly moves from what appears to be the left end of the valley towards the right. The shadow is also moving away from the viewer on an angle that is always gradually increasing relative to the object and the viewer. This is the time when the shadow is young. This is when you can see the wave of the shadow crossing the actual surface of the object as it passes through and over the object's craters and the towering crater rim pieces. This is truly an awesome sight all by itself!

14. The object frames the left side of the view of the valley.

15. The wave of the shadow moves beyond the object's far upper right horizon and into the distance of the light from the Moon that has been transformed into the sight of the valley.

16. Below in the scene on the left, the object's forward moving, forward rolling, slowly sinking surface has moved lower relative to it's original position when the Moon, the valley, was first seen. The light from the object's surface area that bordered with the Moon's light at that earlier point is somehow still there. The rolling surface of the object connects or ties into the the Moon light, the valley itself , through the area of light from the object's surface that remains behind, still there somehow. The effect for the viewer is one were even though the object's surface moves and rolls and looks very different from the valley everything you see from the left, the object, all the way to the far right, the unmoving valley, seems to flow naturally from one to another. A very easy and natural transition between the moving object and the unmoving valley is seen. Everything from the left to the right seems as though it is one expansive place. The overall effect is a very natural looking and very convincing valley scene that extends from the great distance of the view to the right, all the way back to the closer valley scene on the left.

17. During the time that the shadow's great wave travels a group of big main shadows start to separate themselves out. The points of this group of long narrow shadows move away from the viewer towards the distance of the valley to the upper right, further and faster than the leading edge of the shadow's fast moving big black wave.

18. From nearly directly below your eye's very far forward viewing position, three big main shadows emerge from the wide group of now very tall and long narrow individual shadows that have emerged from the shadow's wave.

19. All viewer's need to ultimately locate the big main center shadow that leads the main group of three shadows.

20. Within the three main shadows, the middle shadow is twice as wide as the two outer shadows.

21. The middle shadow is taller and longer than the two outer smaller shadows.

22. This central, very big main shadow is now very rapidly moving into the distance of the valley's emerging newly created floor.

23. The whole idea is to focus on the obelisk shaped point of this now very big black shadow that resembles a very rapidly moving and

growing wide and flat black road.

24. The size and shape of the big main pointy black shadow grows all out of proportion when compared to the size of the object.

25. Defiantly all viewer's now want to be very focused on the big main shadow's point, basically as soon as possible.

26. Once the wave of the shadow has left the surface of the object, all viewer's do want to look at the object's far sloping surface's brilliant and very fantastic shapes that remain above the height of the wave of the shadow. It's a must! There is time for a look at these features that are the light from the tops of standing crater rim pieces that are higher than the height of the thickness of the shadow. A very fantastic sight that along with the rest of it needs to be seen to be believed. There is only a very few short seconds available to the viewer for viewing the object's far slope.

*1. It may be that as the object's surface continues moving across and down the light from previously obscured stars and maybe even planets may suddenly be seen in the area down over the far slope along with the light from the tops of crater rim pieces. In the area of the far slope initially or suddenly later I am not sure which, two large triangle shapes are seem. These two very spectacular attention grabbing sights will instantly be very noticeable and easily discernible to viewer's who are looking in this region once they appear. The triangles are located in the foreground in this region. It could be that these triangle shapes are set back within this region but because of their large size and their extra brilliance I may just remember these shapes as being set forward towards the viewer.

*2. I saw these triangle shapes either just before I looked to the left and then saw over the object's raised feature at the normal shaped surface area beyond or I saw the triangle shapes just after I saw over the large raised feature while I was looking back to the right only seconds before the shadow reaches the ancient place in the light were it sculpts and creates the form of the Awesome Good. My best guess is that these two large triangle shapes are not visible initially once the wave of the shadow has passed over the object's upper right downward sloping horizon, but instead slowly emerge or suddenly appear in this region at a later point. I saw the many shinning shapes that are visible above the thickness of the shadow down over the far slope right after the wave of the shadow had passed off the object and I did not notice the triangle shapes. Perhaps the triangle shapes were actually visible at this early point but they had not grown in size and brilliance and prominence yet. I looked left and right and up and down a number of times so I did not see all of the sights that present themselves and I did not see the every detail or sight that evolves throughout it's total evolution. I have to guess and try to piece things together the best I can.

Basically I have the above sequence of events narrowed down as I describe. I am not completely 100% certain in every area as I try to explain. I was not watching the last return and at the same time taking notes and trying to remember exact details and the timing of the sequence of events somehow knowing that thirty plus years later I wound be trying to recall every exact specific detail along with the timing associated with the sights and events that are seen so that I could then write it all down in a coherent manner with complete 100% accuracy. That sort of task would be very difficult or even impossible for even the most prepared viewer who was even ready to put pen to paper immediately after their view of a return was over, never mind a kid who was taken completely by surprise and sent home stunned and reeling not having any sense of the importance of the sights just witnessed. Consider my circumstances and it should be no surprise that I am not completely 100% certain regarding every detail and the exact order and the exact timing of events.

The main points I mention are very accurate. I understand that future viewers will be easily able to follow the details and the sights and the order of events as I describe them. For me this is easy to understand and I am not at all concerned in regards to the main points I mention in my viewer guides and on this web site. In an effort to be as specific as possible and in an effort to raise other certain points for consideration I have decided to try to expand and discuss possibilities in areas that are in areas that are not 100% clear to me. I know of many points that I could mention similar to the points I am mentioning in regards to the scene down over the object's upper right horizon above the far slope.

In paragraphs; *1. and *2., I am trying to narrow things down so that I can try to mention important points and sights so that future viewers can have a wider understanding if possible. This is especially true in regards to trying to provide future viewer's with indicators that will be a signal to them concerning the amount of time left before the big shadow's point draws and sculpts the sight of the Awesome Good at the end of the Black

Road in and out of the changed light of the Moon.

This is a general viewer's guide. I understand that a viewer of this event could easily get lost in the details that are seen. Many details are seen that I could not ever hope to adequately describe in writing. As always there is always much much more as telescope aided future viewers will be able to clearly see for themselves when they are actually seeing this sea of details for themselves. I try to point towards the biggest main points and I try to touch on other notable points. Overall later after the next Return, people will see and clearly understand that I am only skimming the surface concerning all the sights and all the details that are seen. I am trying to actually complete pages so that I can actually start the process of putting these web pages out there so that people will have heard the idea of what happens during a Return.

Not only is this a starting point for interested persons this is a starting point for me as well. I have never been able to talk about, write about, or think about all of this in one sitting or in just one go. Over time I will continue to chip away adding more and trying to build a bigger clearer picture as I go. I am trying to be accurate and helpful and at the same time I'm trying not to get bogged down spinning my wheels in the details.

These points I am mentioning relate to the fact that I realize that overall there is a very good chance that many people who are viewing a return especially from up close with the use of a small backyard telescope may well find themselves lost in the details. It is obvious to me that I could have easily been looking the wrong way at the wrong things at the all important instant in time when the point of the big main shadow was sculpting the shape and form of the Awesome Good. I could have missed it, even though I was watching the Return. That's actually what very nearly happened to me. I happened to look left in time to see over the raised area of the object itself, and then by complete sheer luck I looked back all the way to the right again and saw the big main shadow traveling away at a point in time approximately not even more than three seconds or so before the point of the big main shadow went up and down and up and down. In the next split second the shadow's point draws the squiggly line as I describe.

It is only by sheer luck that I happen to know that you see over the object's raised area on the left seconds before the shadow strikes on the right. I was totally lost in the details during the time before the shadow suddenly draws the squiggly line. Just by luck I happen to know what happens on the left before the shadow strikes on the right.

Certainly if I had of looked left for even a few seconds longer I would have no idea that looking at the big main shadow's point is the overall big main most important point of it all. If my eye was not looking at the very point of the big main shadow, would I have found myself looking down from a place just above and behind the sight of the back of the Awesome Good? I don't know, maybe. I just don't know because that's not what happened to me. The way I saw in the light and what happened to me is all I have to go on and the only thing that I know with 100% certainty.

As I describe else where, I know that a viewer can join in at a later point and actually see because of what happened to my friend's Dad who also saw. Certainly it's as simple as actually looking but having seen the sight of the back of the Awesome Good from above and behind I clearly understand that this place that you see down from is definitely "the place," were all future viewer's do want to be, plain and simple. All of these separate sights I describe happen quickly and then they are over. My entire view of the Return lasted less than a minute, maybe thirty to forty five seconds from start to finish. Judging the entire length of my view does get into an area that turns into a bit of a blur for me. I am fairly certain that my view was less than one minute I think and at least thirty seconds plus for sure. Judging time is a very difficult thing for me to do with accuracy as I mention elsewhere. All the separate sights I describe take only hand full's of seconds or split seconds. You are going to see a very fast paced series of sights and events. Everything happens very fast and a viewer really does only have one chance to get it right. You don't want to be looking the wrong way at something else when it's finally time to look down from above and behind. All viewers do want to get noticed!

This is why I put such a high priority on the sequence and the timing of the sequence of events. Once people see a return and they have seen the changed light for themselves, they will be able to look back and understand why I have spent so much time going over and over the bit about finding and focusing on the point of the big main shadow and the part about what you will see happening if you are looking the wrong way seconds before the big main shadow strikes in the opposite direction. They will have looked down and seen the back of the Awesome Good and if that's were they were seeing down from in the changed light then they would have been there suddenly seeing the Awesome Good turning and getting up to the left and they would have been there for the Big Noticing! This is a central and a very important part of a return and what happens in the changed light from the Moon.

I am very fortunate to be able to help direct future viewer's to this all important place in time in the Moon's changed light. Thank you for your patience. I hope that many many eye's will all be seeing down from that place in the light the next time it's time to get noticed. Overall if that's the way things happen the next time, then I will know that I have made a difference. I will have done the right thing with what I saw and what I know. Some things worry me greatly and keep me up at night. The points I mention in my viewer guides are definitely not among the things that worry me or the things that cause me to lose sleep. The ancient things that I have been fortunate enough to find and read, although sometimes or rather most times jumbled up, do mention many specific points that I can clearly understand and many specific points that do clearly point to what I consider to be the clear and obvious fact that as far as the parts of a return that I saw are concerned, everything happens the exact same way every time. Perhaps the final evolution of the entire event all the way through to the end of the world of light brought forth the plain may change. In fact this does look like the case however as far as the initial sights and the initial sequence of events is concerned, I definitely have the feeling that the specific sights I saw repeat in exact or in near exact detail. My descriptions of the sights I describe turn out to be accurate and reliable. I know that future viewers may even be surprised at just how accurate my descriptions really are once they have seen the sights I describe.

27. Once the object's normal surface area is suddenly visible once again on the left, over the top of the object's raised surface feature the overhang time has nearly run out.

28. The overhang continues to roll forward and very slowly sink. After the wave of the shadow has left the object's surface, the object's surface area that is beyond the top of the overhang becomes visible. The now once again turning and rolling object's original spherical curving shape is seen again.

29. If the viewer sees the object's normal looking surface area over and beyond the overhang there are only a few seconds left!

30. All viewer's now need to very quickly look to the right. Find and focus on the point of the big main obelisk shaped shadow as soon as possible!

31. Find the big main point and focus on not looking away at all any more. The sight seeing is over and the big main event is about to begin!

32. There is no chance that anyone will actually be able to prepare for the truly awesome shocking sights that are about to be seen. Even though there is little advice that could possible help very much at all I will try to give some simple advice that I understand to be advice that is definitely extremely basic and extremely important.

33. Simply said, look and keep looking. This is something that somehow is not an easy thing to do. The viewer will be facing an intensity that is beyond belief! I understand that there is simply no way that I can possibly convey to others any sort of understanding concerning the very incredible intensity and pressure that you will really feel. Knowing ahead of time that the intense pressure and the feeling that you will feel is not harmful as far as I understand it, may help. I resisted the flow of the pressure or at least I think I resisted the pressure. I don't think that my reaction helped my situation. I just simply had no clue that I could feel anything through my eye all the way in through to the place in my head where light goes if you are seeing it. Certainly I know I was instantly very surprised to know that I was suddenly feeling anything at all never mind feeling the way the light forces itself into you. The feeling of the pressure of the light is simply even incomprehensible. It turns out that light has some sort of pressure or at least this very special changed lasting light has some sort of mysterious and very real pressure or force behind it or with it when you see it and when you see what it suddenly looks like and what it does!

34. This has always been a problem for me to understand. I know this but I have no clue as to how something like this works. I saw light, solid as if a statue and moving much slower than I think that light is suppose to move. Certainly I had no idea at the time that I was even looking at actual light when this was happening. I only figured out that I was looking at light, much later.

35. Does anyone know that light can do these sorts of bizarre things? I have no real true idea maybe, but on the other hand I have a feeling that maybe this is not 100% completely understood or more like 100%

unknown. I do know for sure that very soon all the those big scientist types are going to see and feel the way light moves as if a statue, over there going fast but actually going much slower than light is suppose to go. Do they know this already or is this sort of sight going to be a bit of a shock? I have a funny feeling that even Mr. Einstein would have been sent reeling scurrying back to the drawing board! We shall see! I think that soon it will be time to get out the chalk. Then soon after that all you scientific type folks can reboot your computers with all the new things you suddenly were forced to know!

36. Multiply these sorts of surprising types of difficulties with the impossible way that the light actually appears and the way it looks at you face to face and suddenly you have arrived at a place that is it's own extreme shocking reality that grabs you and forces you to know and realize what you are seeing! Impossible contradictions that are real and staring at you, overwhelm and constantly overpower you again and again with every new stunning sight! I really hope that one day I get to sit around with a bunch of other people who have also seen these exact same sights. I am very curious to know what sorts of opinions and experiences and feelings will be offered up and going around the room. Suddenly I won't feel alone anymore. I will feel much better.

37. The entire point is to see the Changed and Lasting Light sights. All viewer's will need to continually refocus on the idea of not looking away so that they will be able to remain actually looking. Brace yourself and try your best. If you don't keep looking you will regret it later.

38. Focus on the very point of the big main shadow. Allow your eye to follow and ride this most distant and central main shadow's point's movement and you will find your eye's viewing position to be magically drawn always farther forward into the distance of the light of the Moon.

39. Very suddenly the point starts to very rapidly go up and down and up and down. At this point the big main shadow reminded me of a flat ribbon as it briefly rapidly when up and down and up and down.

40. Almost instantly, or instantly after the shadow's up and down motion stops, the cutting left and right zig zagging squiggly line starts to happen!

41. If you are looking, you have arrived! The intensity is yet to strike but down below and in front of were you are seeing down from the sight of the very muscular back of the Awesome Good is there!

42. Your eye's viewing position is from just above and behind. It is as though you are looking down from a height of about 15 - 20 ft. Maybe slightly higher but from no more than 25 ft. (The size and power of the telescope you are using may be a factor to consider here.) A second or two or maybe three and then suddenly the ultimate instant in time is about to happen next! But just before the ultimate instant happens the lasting light moves up. The sight of the back of the Awesome Good drifts up closer when the light moves up. Suddenly a flat smooth place or area is seen surrounding the Awesome Good.

43. In between the light's upward motion the light somehow magically builds outwards turning the flat white cloudy effect that surrounded the Awesome Good into what looks like flat hard ground. You can see right out to the edge of the flat smooth area. Beyond the edge of the flat smooth floor or ground area where the Awesome Good is now seen positioned you are able to see the indiscernible look of the flat white cloud effect again. This is true out to the left of were the Awesome Good is positioned. I did look out to the edge of this sudden new flat area to the left of the Awesome Good and so I know that there is an "edge,' to be seen in this one spot. I don't know if there was any sort of discernible "edge to be seen anywhere else radiating out from the Awesome Good's central position. Flat ground did surround the the spot were the Awesome Good is situated at this point and if you look left you will see the edge of the new flat smooth area and past to the flat white cloud effect.

44. The changed and now sculpted incredible solid lasting light from the Moon, the sight of the back of the Awesome Good, is very clearly there! The viewer has only enough time to realize what they are seeing before the impossible Awesome Good that they have suddenly found themselves looking at, suddenly gets up, and turns to the left all in one fast spinning motion, and turns to face you and look back up directly at you!

45. It turns out there really is an ancient Big Thing and this is it!
Instantly at this same incredible instant the beginning of the completely overpowering mind bending intensity is nearing and building and getting ready to strike! An incredible thing is happening! The way this light forces itself is at the heart of the question of human life itself. If you are looking you are made to see and comprehend. The shape of the light that has focused on you must have been first. Humans are here on Earth! How does this work? How could a situation like this exist? I certainly don't know but somehow this situation does exist.

46. In this now completely impossible situation the simple advice I offered concerning the idea of remembering to try to focus on the idea of not looking away will be at best, only a very very small help. Remembering that you are suppose to look and you are suppose to look for as long as possible is a good place to start.

47. Also it is plainly obvious to me that the changed light from the Moon is beneficial and good for us. It must be. I don't know how it does what it does but this fantastic changed light does definitely have very special qualities.

48. The view of the valley, phase one of the illusion ends with the Big Noticing. Now very suddenly a short transition period takes place before phase two of the illusion, the mountain, begins.

49. During this transition time period the view of the valley transforms downward in in size. It is still there but scaled down or focused inwards. The first phase of the lasting light event the valley is over. The transition period with it's repeating inward pulses ends.

50. Now the second phase of the lasting light event has started. The light lasts and rises as more of it arrives from over the object's upper right horizon. The light from the Moon arrives from over the object's upper right horizon as one way arriving and then slowed down areas or waves of basically moving as one areas of almost solid moving light. Incredibly viewers are going to literally see rising and lasting real light from the real rock of the Moon. A very convincing looking mountain is very suddenly rising up through space growing at it's base complete with the valley contained down within the upper structure that is different from the bending curving column of mountain that has formed below. Once the one way moving areas of light have joined in at the base that is the point when the light, that is already effected and changed, actually becomes "solid and lasting."

51. A completely incredible totally stunning and because of the intense pressure that is felt, an even overpowering sight! Concerning the upper structure itself, I have seen pictures of ancient Maya buildings that are exact copy's of the way that the upper structure looks. Certainly the entire bending curvingly straight height of the mountain of light is not constructed but the mountain's upper structure itself is clearly the focus of the builders. Definitely this entire event was very much the focus of the ancient Maya, this is plainly obvious to me. In fact one day it will not

just be me but everyone is going to realize that a great majority of ancient constructions are in fact models of the way the lasting light looks. I know that this can only sound impossible but if you have ever seen the lasting light, you will have been forced to know that this is simply the way it is. It turns out that as usual it's almost always about the lasting light and the way it looks at you and the way it appears and the very bizarre strange way it behaves.

52. The lasting light mountain is rapidly rising and the face to face ultimate staring contest is still on and has not been interrupted by the transition period's inward pulsing or by the way the Actual Awesome Good itself has changed or evolved in appearance during the transition period. Now at this point the full force and intense pressure is instantly suddenly being felt and literally received by the viewer.

I realize this can only sound impossible, I agree. Anyways I can't let what should be impossible be any sort of factor for me in my decision to try to describe my view in an effort to provide future viewer's accurate information, although it's not easy to ignore these sorts of problems that I face. The impossible is going to be happening again and I realize that I obviously can't change or effect how the light looks. What am I suppose to do? Not describe what the lasting light looks like because of what and how people are going to think? At one time I used to beat around the bush when I wrote, now I don't. It is what it is and the light does what it does. It does not matter how impossible this sounds to people. It turns out that the opinions that will be expressed against my descriptions are completely totally irrelevant. I don't mean to sound rude towards people and it's not as if I don't respect the opinions of others in fact I do have a tremendous amount of respect for the opinions of others. As it happens this has nothing to do with what I think or what anyone else thinks and I am able to realize this fact. If a person can't stop to hear the idea of a return then that's fine. I'm not trying to convince anyone, that is not a thing that is possible for me to do. If you want to know a bit about what the changed light looks like then read on. I am not try to actually get you or even interested persons to believe my descriptions. I am offering you a chance to read and hear about what happens during a return. I did see the big thing that happens between the Earth and Moon and I am trying my best to describe these particular very strange and totally bizarre sights. I am doing this because it's the right thing to do. The sights I describe and the points I mention are accurate and genuine.

53. As each wave or area of moving light arrives from over the object's

upper right horizon and then moves and forms the next area and solid moving light, it has joined in the base of the mountain and is itself, the newest area of mountain rising. Once the areas of one way moving pulsing forward waves of light have arrived and have finished moving as it's own area of moving light and has become the newest area of rising mountain they remain unchanging. It's the light from the rock of the Moon. You are seeing it but it's this great big huge thing rising over there. A very very convincing mountain is seen although you don't normally see mountains rising bending curving up through space growing at their base. It is what it is and it does look like rock or more like smooth stone. In fact as a kid for some years I thought of it as a smooth sided stone mountain not a rock mountain which I suppose I was thinking would look rougher. As far as the very strange waves that arrive at the base and become new mountain rising I was not really able to think about this very much. I knew the mountain was somehow magical but for a long time I had no idea that it is actually light. Even once I finally figured out that the telescope had not been moved I still had no idea that the magic stone mountain was light from the Moon.

54. What happens when you are actually looking at something, say a tree? For me, I don't automatically think to myself, ah ha, there's light from at tree. Instead I simply see a tree. That's what it is, it's a tree, not light. If you see the Magic Light Mountain, that's what it is, a mountain. A strange mountain, but you see it the way that it is and that's what it looks like. When I suddenly saw the Magic Light "stone," mountain, that's what I saw, a mountain. I didn't realize anything more. This was a very strange mountain, but it was a mountain. Even though it does this bizarre thing and it has the Awesome Good along with the other forms sculpted by shadows located down within the upper structure with the Awesome Good, it looks like a familiar thing. It looks like a mountain that is as solid as stone.

It turns out that in many ways the mountain is light from solid stone or at least from something solid, the Moon. The Moon is real. When you are looking at the changed light mountain you are looking at real light from a real place. In spite of the bizarre nature of the changed light event it turns out that in the area concerning the mountain's appearance, it should be no surprise that the mountain itself looks real. It does look like stone or rock as some others in the past have also described the mountain. This is an unbelievable thing I know but I understand that future viewer's will benefit if they have some idea what they are seeing when they are seeing it. The fantastic sights that are seen above the object's forward rolling surface are all made of light. It's all real light from real places mostly the Moon but perhaps a number of stars as well and maybe even planets I don't know but on the most basic overall fundamental level it turns out that the vast majority of all of the incredible sights that the viewer sees above the rolling surface is light from the Moon after it has been somehow changed by mysterious power of the big rolling place that returns and crosses between the Earth and the Moon.

55. Certainly this does sound impossible but it turns out that this is what really happens. Ultimately a situation is developing and evolving were below you can clearly see the turning rolling crossing slowly sinking disc shaped spherical object. Above the object, the lasting light mountain, tall and narrow curves to the left back over and against the object's left to right motion and then, "ultimately," the oval shaped "plain," the surface of the Moon complete with the forms sculpted by the points of all the shadows, emerges out from down inside the upper structure of the lasting light mountain. At least, up to whatever exact point in history, the oval shaped area of lasting light emerged from down inside the top of the mountain of lasting light. This is exactly what use to happen and that's that, as they say.

56. Then the formerly curving straight light mountain turns into a snake like line and the object separates from in front of the Moon falling away down further towards the lower right. Check "The Great Serpent Mound," located in Adams County, Ohio, U.S.A. Ultimately that's the scene. My view ended while the light was still building at it's base. I did not see the oval shaped light from the surface of the Moon emerge from down inside the lasting light mountain's upper structure. I did not see the eclipse actually end. The Serpent mound shows the changed light no longer emanating from over the object with the Moon partially behind it the way it was doing so during the eclipse. After the separation the light has not finished emanating from over the object. The spiral effect is the rolling motion of the object with the light still emanating from over it's surface.

57. The area that was originally the flat cloudy area within the valley's Mountains becomes, is the plain or the firmament the oval shaped area. The mountains that were originally seen surrounding the flat white cloud of the valley became focused inwards and became the walls of the sunken room. Later these same mountains become a sort of sleeve of lasting light and the outer walls or area of the rising lasting light mountain. Later after the oval shaped area of changed lasting Moon light

emerges out from down inside the tall narrow bending curvingly straight mountain, the mountain itself, the sleeve of outer light that contained and stored the oval shaped area of lasting light, is straight no more and it falls away down to the right or it just becomes a squiggly curving snake like line. Then the oval shaped plain continues towards the Earth and is described in ancient texts and it the focus of the ancient texts.

58. The original face to face staring contest starts when the Awesome Good turns around to the left and stands up to directly face you. During the rising of the growing mountain you are seeing from a position that is back much closer to a place that is much more like what you would expect a normal viewing position or perspective to be like. From this much more normal position there is much more distance between your position and the position of the rapidly rising nearing position of the Awesome Good compared to the face to face moment. This is true but at the same time with the sudden pressure and screaming intensity that is suddenly felt and known by the viewer, the ultimate staring contest is on even more than ever. The impossible Awesome Good is still there going face to face with you from his central and forward prominent position.

59. Suddenly it is clearly obvious that the focal point of the Changed Light is still the focal point, and it is still defiantly squarely focused on you, but it looks different. It has changed! It's the second version of the Man of Light! It transformed during the transition time period and it now has a whole new look, although it's focus has retained it's original penetrating stare! An incredible thing to see!

60. The object's upper right horizon was and is as if a very thick lid being drawn slowly back open showing more as the sunken room rises. The valley area, the surface of the moon is seen reduced in size, now the sunken room now down inside the top of the mountain as the mountain continues rising bending curving very directly very rapidly directly at the viewer.

61. Down inside the top of the rapidly rising mountain of Moon light, the Awesome Good is closing towards you. Your eye's viewing position was returned back to what you would think of as being a more normal and far back viewing position during the transition time period between the valley and the beginning of the rising mountain.

62. Incredibly the telescope added viewer is going to see that this intensely focused light is rapidly nearing. Incredibly down inside all of

this, this now huge growing thing the Awesome Good, the focal point of the entire event itself, with great speed and momentum is very rapidly closing on your position!

63. Down inside the mountain's upper structure the valley now looks exactly like a sunken room. There is an array of separate forms located around the focal point. All of these separate forms the main elements are created by the other main shadows that reached further faster than the shadow's main wave but after the main shadow sculpted the form of the Awesome Good. Just before my view ended I looked away from the Awesome Good and to the left and slightly up around the curve of the shape of the light from the curve of the object's upper right horizon at the next human shaped form.

64. The next form over in the direction I describe may have been created by the shadow that traveled to the left and slightly behind the big main shadow or it could have been created by the shadow that was the next main shadow further to the left, I'm not sure which. I am guessing, but for various reasons, I am leaning towards the idea that "the next main shadow further to the left," is responsible for sculpting this form that I am referring too. I looked at the top of this form and then down the front of this form to the valley floor below, now seen as the floor of the sunken room.

65. My eye went from the bottom of the front of this form across the floor traveling left in the view finder to the base of the sunken room's wall, and then up and out of the inside of the sunken room to the outside, outer wall of the sunken room which is actually the outside of the top of the rising mountain, the upper structure. Again my eye traveled from the front of the base of this second form that was created in the valley to the left towards rooms wall. What is seen as the walls of the sunken room are actually the steep faced mountains that were originally seen ringing the valley. The sight of the sunken room is actually the valley rising reduced in scale with the forms that were created by the shadows, contained and arrayed within the floor area of the sunken room. A very incredible thing to see.

66. Once my eye was outside, I saw how the upper structure looked and I saw how the rest of the mountain looked as it curved back down to it's base. I looked all the way back down down to the mountain's base, then to the object's turning surface, now very much wheel like. The last few seconds of my view were spent watching the newly arriving areas of

slowed down and moving areas of light smoothly and naturally joining in and becoming the base of the mountain as new areas of created and unchanging areas of moving rising mountain. These one way areas of moving light pulsed forward and joined in as new mountain with a pulsing motion or movement that is very similar to the way the the transition period's inward focusing motion pulsed inwards. I can guarantee that future viewer's will be totally amazed by even just the sight of the bottom of the mountain building alone. All by itself the the sight of the way that the light arrives and joins in with everything that is already there, creating and storing more layers of light is truly an fantastic spectacle!

67. This action is actually central to the entire lasting light event. These new pulsing one way waves that arrive and then join into the base of the mountain are like the next new frames in a long stripe of movie film footage. Each forward pulse is it's own layer in the mountain and another picture in a continuous movie that used to be seen later as a part of and within the plain. Also each separate pulse or wave or layer in the mountain records and stores a time in the light with the continuously evolving influences of the shadows cast off the moving rolling object's crater rim pieces into the light from the Moon. Each one way pulsing area of light is a type movie onto itself.

68. The entire scene on the plain and all the forms that are seen on the plain are seen in action later by viewer's on Earth as if they were a part of a fantastic movie. The light records and stores transports and then plays or replays all of the results of the shadows sculpting and creating in the light that happened earlier when the shadows were actually being cast off the object into the changed light from the Moon. As the one way moving areas or ways emerge from over the object they contain everything that just happened and they contain everything that is seen happening later.

69. As usual this is to bizarre to explain in words or even simply to bizarre to understand at all. I am able to read ancient texts that describe the way the light arrives and forms or reforms or is presented when the plain emerges and is presented. I am able to read on through these descriptions of what the lasting light does next. I saw the first part of the sculpting in the light and I am able to read about what happens next starting with chapter one. This is truly a stunning overwhelming thing to realize. I am always in some sort of shock because of what I saw and as I realize more I am always only more overwhelmed. Like I said, it's plainly obvious to me that I saw the big thing that happens. Also it's

plainly obvious to me that this big thing that happens is about to happen again, very shortly. I know this is impossible for future viewer's as well as for everyone else but BE READY!

70. I don't know what be ready means but that's all I can think of to write at this point. I know we have to be ready but how are we suppose to actually do that? We need to see the object returning before it gets here but is it possible to actually be ready? I don't know. These types of questions and the answers to these types of questions are way to difficult for me. I am able to realize that these sorts of questions need to be asked and they need to be answered. This is obvious to me.

71. Then, unfortunately my view ended. I describe and discuss this point in detail elsewhere.

72. Will the entire evolving scene remain within the telescope's view finder right through the entire event including phase three of the Changed Light event I don't know. I think that having binoculars available may prove to be a good idea. Along with possibly enabling the viewer to step back a bit in order to get back from the big sight, binoculars may provide a degree of aiming flexibility that may or may not be required.

73. I think that part of the point that I am trying to make at this point is that viewer's should try to be ready for the next sudden unexpected evolution of the lasting light event. Later during phase three of the lasting light event, the oval shaped plain the light reorients itself somehow and this causes your eye's viewing perspective to change. From what I have seen from looking at ancient depictions the viewing perspective or angle that the viewer has when looking at the valley is the same or very similar to the angle that the viewer has when they see phase three of the Changed Light event, the plain.

74. First your eye's viewing position moves forward drawn closer. Then your eye's viewing perspective is from farther back as you watch the mountain rise. Later during phase three the plain when the original sight of the valley is seen again, minus the original surrounding mountains, from up close, it may be possible to see the entire event through a telescope without the need to switch to the lower magnification and wider field of view that would be provided by binoculars. On the other hand, maybe not. This gets into unknown territory for me. Maybe the naked eye will be all that's required just like in ancient times.

75. If I saw it I know it and I can describe it. As far as the rest of the visual event is concerned I have to guess based on what I read and what I see in ancient depictions. It looks very much like a small backyard telescope will work just fine right up to the end of phase two of the illusion, the rising mountain of light, or nearly up to the end of the second phase of the Changed and Lasting Light event. Certainly the view through a small back yard telescope would have worked out just fine for some time past the end of my view that night long ago. Exactly how far past the end of my view I can really only guess. Definitely in ancient times when the third phase of the Changed Light event occurred and the Moon's changed light arrived at whatever height above the Earth a telescope was simply not required and obviously was not available. Also when the light arrives here a telescope may become completely useless. Binoculars may prove to be worthwhile along with the naked eye at this point. Time will or it won't decide this one.

76. It turns out that it's all about the Moon's changed light from the start to the spectacular finish. Will the entire lasting light event occur and play out in it's full entirety? Will we see the world of light arrive and then age and then end? Will we see the end of the world of lasting light just like the ancient people saw and describe? Even if we don't see the sight of the lasting light reach all the way to the end of it's fullest longest potential time period along with the entire event from start to finish like it used to do in ancient times, everything we will see will be spectacular!

77. We have a chance to see and know what our most distant ancestors saw and knew. It turns out that in some ways, in some areas, our distant ancestors did have a better understanding about what goes on around us in our world and in our own local area of space. They knew this fantastic thing and we don't. I am sure that our ancestors never could have imagined that knowledge of something so big and so fundamental to everything could somehow one day be basically lost. Listen to how they wrote. The audience of the day simply knew what goes on plain and simple. It seems the authors of the day never stopped and turned around and when back to square one in an effort to take it from the top for the benefit of those who did not know. Everyone knew and it's obvious that many ancient people did have a deep and very sophisticated knowledge and understanding concerning the object and the details within the lasting light and we don't.

78. It turns out that there is an orbiting returning massive crater covered, forward rolling incredible moon type of moon coloured object and it's

returning and it's just about to roll by again, plain and simple. Ready or not! There is no stopping this from happening. This is going to happen again as always soon.

79. Fortunately whatever the down side if there is a down side the sight of the ancient mystery does happen over the object's upper right horizon in the fantastically changed light from the Moon. Incredibly it's all about light that has been changed by the forces and raw power of this truly fantastic object!

80. I say Fortunately. I say Fortunately because I am glad that there really is a big thing that happens. The Ancient Object can cross down in front of the Moon safely and is a very fantastic thing! Plain and simple!
It turns out that the ancient texts describe an event that really happens. Many people are convinced that the ancient texts they follow are based on real events. This is what I see around me and this is what I hear around me. It turns out that the ancient texts actually really are based on a real event. This is a fantastic thing at least in my opinion this is a fantastic thing. People are all entitled to their own opinion and so am I.
The ancient object itself is beyond spectacular and very incredible! The way the object's powerful forces change the light from the Moon is simply the most fantastic incredible thing that happens anywhere, anytime!

Good Luck and remember there is such a thing as a safe Return.
The Ancient Object's last return orbit down was a safe Return!

All the best to you and yours,
Thanks... D.S.W.

ABOUT THE AUTHOR

I was born in Montreal Quebec, Canada in 1960. Our family moved to Mississauga Ontario in 1965. I am very fortunate to be a father and very proud of my two children and I love them very much. Together with my very special Mom and Dad and my two wonderful sisters and our entire family my life has been full of love and complete.

www.ingramcontent.com/pod-product-compliance
Lightning Source LLC
Chambersburg PA
CBHW020909180526
45163CB00007B/2679